IN PRAISE OF FLOODS

In Praise of Floods

The Untamed River and the Life It Brings

JAMES C. SCOTT

Yale UNIVERSITY PRESS
New Haven and London

Published with assistance from the foundation established in memory
of Amasa Stone Mather of the Class of 1907, Yale College.

Copyright © 2025 by Yale University.
All rights reserved.
This book may not be reproduced, in whole or in part, including illustrations,
in any form (beyond that copying permitted by Sections 107 and 108 of the U.S.
Copyright Law and except by reviewers for the public press), without written
permission from the publishers.

Yale University Press books may be purchased in quantity for
educational, business, or promotional use. For information,
please e-mail sales.press@yale.edu (U.S. office) or
sales@yaleup.co.uk (U.K. office).

Ko Ko Thett, "The Chindwin [River rage]" from *Bamboophobia*.
Copyright © 2022 by Ko Ko Thett. Reprinted with the permission of
The Permissions Company, LLC on behalf of Zephyr Press, zephyrpress.org.

Designed by Mary Valencia.
Set in Minion Pro type by Westchester Publishing Services.
Printed in the United States of America.

Library of Congress Control Number: 2024937393
ISBN 978-0-300-27849-1 (hardcover : alk. paper)

A catalogue record for this book is available from the British Library.

This paper meets the requirements of ANSI/NISO Z39.48-1992
(Permanence of Paper).

10 9 8 7 6 5 4 3 2 1

Dedicated to those who revere rivers, the creatures for whom a river is their lifeworld, the Burmese peoples, and especially Maung Maung Oo and Naing Tun Lin

The Chindwin [River rage]

She rises above the flood stage like an overfilled pot—a tight pot that doesn't leak.
Hiccupping like a soon-to-be single mother, she will puke back into your face all the plastic and rubber you've forced into her throat. She belches stale draft. She is a river—hanging on a river hangover.

Rice hoarders will be whipped. Split bean hoarders will be spared. She will show what a dominatrix she is to those who mistake the rivulet Mu for a river. Cross with the land, she will piss on the road's shoulder.

Her refuse will fill disaster relief bowls. For her monthly does she have to know the day of the calendar month? If there's no bloody drought there will be a bloody deluge.

Local poets no longer make a distinction between
[river water] and [tear].
And now, how will you unfuck her?

—Ko Ko Thett

ချင်းတွင်း [မြစ်ရဲ့ အမျက်]

ဆီလို အပေါက်ရှာတာ မဟုတ် အပြည့်လွန်အိုးမို့
လျှံတက်တာ လင်ကောင်မပေါ် တဲ့ဗိုက်က

ကြို့ထိုးတာ မဟုတ် မြစ်နာကျလို့ လည်ချောင်းထဲက
ကြွတ်ကြွတ်အိတ်တွေကို ဝေ့ ဝေ့ နဲ့ ပြန် အန်ထုတ်တာ

ပဲမလှောင်ပဲ ဆန်လှောင်တဲ့သူတွေကို ရွှေး နှိပ်စက်
မူးမြစ်ထင်သူတွေကို အမောက်ထောင်ပြ လမ်းဘေး

ချိုးပါနေတဲ့ နိုင်ငံတော်ကို မေးငေါ် မယ်ဘော်က ကဲကဲ
တွေထက် ကပြတာ သူမ အန်ဖတ်တွေ ရေဘေး

အလှူခံခွက်တွေ ပြည့်တဲ့အထိ ပြက္ခဒိန်ကြည့်ပြီးမှ
ရာသီလာရမှာလား မိုးခေါင်ရင်ခေါင် မိုးမခေါင်ရင်

ရေကြီးမှာပဲ သူမကြောင့် ဒေသခံကဗျာဆရာတွေ
မြစ်ရေနဲ့ မျက်ရည် မသဲကွဲတော့တာ

CONTENTS

Preface ix

Acknowledgments xix

Introduction: A Word about Rivers 1

ONE Rivers: Time and Motion 8

TWO In Praise of Floods: Moving with the River 38

THREE Agriculture and Rivers: A Long History 64

INTERLUDE An Introduction to the Ayeyarwady 77

FOUR Intervention 107

FIVE Nonhuman Species 133

SIX Iatrogenic Effects 153

Notes 189

Illustration Credits 209

Index 213

PREFACE

I've thought about rivers and streams for a long time, though only recently in a serious, scholarly way. Before, my acquaintance was both casual and recreational. Growing up along the Delaware River when it was at its most polluted, twenty miles upstream from Philadelphia on the New Jersey side, I swam and fished its waters for eels. Later, with two high school friends, I canoed the upper Potomac for several days until we capsized and lost most of our gear on the bank opposite Harper's Ferry. Our accident took on a lengthy afterlife in my family's lore. While we scrambled to hitchhike to the cars we had left at either end of the planned route, I didn't think to call home. As luck, both bad and good, would have it, a local fisher angling from a flat-bottomed boat happened to hook onto my submerged blue jeans containing my wallet and driver's license. Imagining the worst, he found my home phone number and called to break the news about what he had found. Having heard nothing from me or my companions, my mother assumed the worst about the fate of her youngest son. She suffered two days of panic followed by grief before I innocently called home. My mother never forgave me for what I had put her through.

Much later, I spent six or seven consecutive summers living with my own family in a rustic hunting cabin on the bank of Penn's Creek, a well-known trout stream in central Pennsylvania. Instructed by my extended family and local subsistence hunters and fishers, I moved from rank amateur to proud mediocrity in my fishing and canoeing skills.

Never, in the course of my casual acquaintance with rivers and streams, did I imagine that I would one day presume to teach about rivers, let alone dare to write about them. That I could safely leave in the hands of brilliant authors such as John McPhee, Wallace Stegner, Ellen Wohl, and Mark Twain.

My long-standing interest in rivers and my scholarly life came together in a thoroughly opportunistic fashion. Having spent a mesmerizing year in Burma well before I began graduate school, I yearned to make it my terrain of research. But it was essentially off limits to any such endeavors from 1962 until 2011. When I finally returned, I was eager to throw myself at the language and see as much of the country as possible. Travel within Burma was still restricted, and I was frequently stopped at police and army checkpoints and asked for my passport and visa and to explain my presence. I was often followed. Since I had traveled a good deal of the country by boat along the Ayeyarwady, and since this great river, along with its tributaries (and distributaries), covers much of the country, I hit upon it as a convenient and plausible explanation for my presence. Whenever I was stopped, chances were that it would be close to the Ayeyarwady or one of its branches. I would simply say that I was studying the Ayeyarwady and, for the most part, this tactic worked. What began largely as a deception gradually became a preoccupation as, willy-nilly, I learned more and more about the river.

When it eventually dawned on me that I might actually try to learn more about rivers by teaching an undergraduate seminar on the subject, there were at least two nagging questions that had been bedeviling me about rivers, each related to my casual acquaintance with them, that I had to solve first.

The first arose from an amiable conversation suddenly gone awry with a hydrologist. It occurred at a residential conference site where two meetings were being convened in the late 1970s—one for Southeast Asianists like myself and another for engineer-hydrologists. As we dined together at lunch and dinner, we were urged by our hosts to get to know one another and converse across our narrow specializations. Taking this large-spirited advice to heart, I found myself sitting next to a widely read Filipino hydrologist on the third evening of the conferences. Casting about for an opening, I recalled having learned in the previous year that the Colorado River, diverted and impounded by dams, never reached the Sea of Cortez for much of the year. The fact itself had stuck with me and triggered a distinct sense of sadness on behalf of a river that was "denied" its natural destiny: to flow into the sea.

So, in my effort to please my hosts, I told him what I had learned about the Colorado River and asked, "Wasn't it sad, given all our poems about rivers running down to the sea, that the Colorado was prevented from achieving its destiny?" He abruptly put down his fork, turned to face me directly, and declared: "No, no, no! It is not a sad thing at all! It is wonderful; it means that all the water in the Colorado is used for important human purposes and not a drop is wasted!" That the Colorado didn't get to the sea, he believed, should be a source of jubilation rather than sadness. I realized, then and there, that we would not have a long conversation.

This encounter was, in its way, diagnostic; it sprang to mind again and again as I read through the literature on rivers. My Filipino engineer was a perfect example of the sort of utilitarianism that views nature as "natural resources"—as a factor of production in the satisfaction of needs—most specifically the needs of a single species, *Homo sapiens*. He was hardly alone in his conviction; his view was, at the time, hegemonic. It was what one would have expected from the American Bureau of Reclamation or the Army Corps of Engineers. In speaking of the storied Nile River, even Winston Churchill echoed this idea (as quoted by Patrick McCully in *Silenced Rivers*), though more lyrically than my Filipino hydrologist: "One day, every last drop of water that drains into the whole valley of the Nile shall be equally and amicably divided among the river people and the Nile itself . . . shall perish gloriously and never reach the sea."

Joseph Stalin, whose aspirations for reengineering rivers were far more expansive than Churchill's, shared his views (as quoted in Steven Solomon's *Water: The Epic Struggle for Wealth, Power and Civilization*), if in more prosaic terms: "Water which is allowed to enter the sea is wasted."

What is notable here and what will preoccupy us in this long essay on rivers is the way in which, for Churchill and Stalin, there were only two variables at play. The river is reduced to water, just so much H_2O that must be divided between rival claimants. And the claimants? They are, in turn, reduced to us—*Homo sapiens*. Gone are all the other beings in and around the river for whom the waterway is their indispensable lifeworld; fish, amphibians, reptiles, shellfish, water birds, wading birds, raptors, riverine mammals, insects, microbial life, and algae. Gone from the water itself is what it carries: silt, soil, gravel, clay, sand, and organic matter that, if left to its own devices, will be distributed along the floodplain as the river wends its way to the sea. Cost-benefit calculations are so much easier when there are only two variables to consider: water and humans. This book is motivated by my effort to understand what happens when humans endeavor to "tame" the river, to sculpt and script its movements to serve the (short-term) interest of humankind alone. What we have done to rivers and the consequences of our actions seems a powerful metaphor for understanding the troubled and possibly cataclysmic relationship between humans and nature generally.

The second question that dogged my thoughts arose from what I thought I knew about the movement of rivers from nearly a decade of summers spent along Penn's Creek. The old hunting cabin in which we lived was within thirty feet of the creek and, though raised a bit off the ground by stone slabs, not more than a couple of feet above the average spring high-water levels. A visit in February after a quick thaw would often reveal broken ice floes leaning against the upstream side of the cabin. The force of the current was gradually eating away at the bank near the cabin. Over the previous decade, the early spring current had swept away a large oak stump that had been used to delineate the western extremity of the property. Observing this process near the cabin and at other bends and meanders where I fished year after year, I thought of the stream as an ever-moving, ever-changing process defying our sense of immobility as implied by mapmaking. It was, I thought, a gradual process as the creek, little by little, year by year, carved out a slightly new course, thus remaking the landscape. (As an amateur hydrologist myself, I spent part of one summer piling up rocks from the streambed to build a small diversion upstream from the cabin to redirect some of the flow away from the eroding bank. The flood of 1972 instantly erased any trace of my embankment.)

My gradualist view of stream morphology was abruptly destroyed near the end of June 1972 by a huge flood that submerged the first floor of the cabin and knocked out an old bridge a few hundred feet downstream. In several places, thanks to Hurricane Agnes, the stream had jumped its banks and carved out an entirely new channel. This was not gradualism; this was sudden and explosive. I realized then that most of the consequential changes in Penn's Creek channel over the past several decades had occurred in a few short hours of rampaging high water at the peak of the flood. The gradual change I had observed over the years was, by comparison, on a long view, trivial; much of the channel was obliterated in these few short hours. The creek was moving, alright, but to understand the nature of that movement I had to open the temporal lens far wider than a decade of summers. Even my perception of gradual change was defective. My time along the creek was mostly from June through August, when water levels are lower than average and variability modest. What I had discerned as gradual change year by year was probably accomplished suddenly in the few hours, typically in mid- to late February, when snowmelt, a good rain, and broken-up ice floes created an annual mini-flood stage. The changes I had interpreted as "accretion" were, in fact, largely the result of what is technically known as "avulsion." The two are treated differently in international law and property law. Thus, when a jurisdictional boundary is demarcated by a river channel, and when the river channel gradually shifts by accretion, the boundary moves with the river. If, however, the river channel makes a sudden and decisive change—say, in a great flood surge—the law provides that the boundary remains defined by the old, now-abandoned river channel. (One imagines thousands of court cases in which the precise line between accretion and avulsion is the point of contention.) On the widest possible temporal view, of course, many of the greatest changes in rivers have been the result of geological events.

My fascination with Burma and its people, cultures, and landscapes, not to mention its main river the Ayeyarwady, predates my graduate training. A series of improbable mistakes and coincidences led me to write a senior thesis on the political economy of Burma. Until then, I was headed to law school, in part as a way of postponing more confining career choices. On a whim, I applied for a Rotary International Fellowship to Burma, and, to my astonishment, was selected.

The academic years 1958–60 in Burma, at Rangoon University and then Mandalay University, were punctuated by long trips with Burmese friends throughout the country by motorcycle—in my case, a decrepit 1940 Triumph. The friendships and experiences from that year changed my life. I switched from law school to political science and began to think of myself as a specialist in Burma and Southeast Asia. When the time came to pick a thesis topic, the Ne Win military regime had all but prohibited research by foreigners. By default, I became a Malaysian specialist and would subsequently spend a year and a half there living in and studying a rice-farming village.

It was only in the early days of the 2000s that travel requirements to Burma were relaxed, and only from 2010 to 2020 when, thanks to Aung San Suu Kyi and power sharing, the country became a relatively open society. Seizing the opportunity of pursuing my first love, I returned regularly and worked to hone my Burmese speaking, reading, and writing skills.

Alas, as I write this in November 2023, the curtain has fallen on that brief decade of opening since the junta annulled the legitimate elections of 2020 and seized absolute power on February 1, 2021. The country has suffered killings, detentions, bombings, arson, and brutal repression. Resistance by ethnic armed groups (themselves no strangers to military repression), and now by huge segments of the general Burman population, has resulted in a bloody stalemate. As a political scientist and student of resistance, I don't believe I have ever encountered a civil war over democracy and federalism of this magnitude. Essentially, the entire civil society is arrayed, passively or militantly, against a cocooned military with its own schools, hospitals, supply chains, and pensions. The military has no legitimacy among the civilian population. Soldiers once proud to wear their uniforms in public now don civilian dress to avoid the hostile stares and comments of the population. The army, or Sit-Tat, apart from their allied and subsidized militias, is encouraged to plunder and burn villages. Regardless of the eventual outcome, this democratic rebellion of the civil society as a whole against an isolated military will find a distinctive and diagnostic place in the study of revolutionary movements.

The reader may notice that I usually employ the name Burma instead of Myanmar. Both terms refer to the same nation-state. The reason is baldly political and, alas, entirely performative. Myanmar came into official use as a way of whitewashing the badly tarnished reputation of military rule in "Burma." (Recall how the "Congo" became "Zaire" and later the "Democratic Republic of the Congo"!)

The book before you contains more disjunctions than I would have preferred. They are due to a perfect storm of research complications. Because of the violence unleashed by the coup, trips and interviews I had contemplated conducting along the Ayeyarwady River were suddenly impossible. Even prior to the junta coup of early 2021, the COVID pandemic precluded travel to Burma for the Ayeyarwady-related research I had planned. Once the coup occurred, I, along with many others openly associated with the democratic opposition, were barred from travel to, let alone cleared for research in, Burma. My role in establishing Mutualaidmyanmar, a charitable website designed to support civil servants engaged in peaceful resistance, together with my public appearances and published writing, made it inconceivable that I would be permitted to return as long as the junta was still in power. Even in the absence of these insurmountable barriers, my own age-related infirmities would have made the research I envisioned difficult if not impossible.

Several of the commentators evaluating this book in manuscript noticed that the section on river spirits (*nats*) and the much larger section describing the eco zones, hydrology of the Ayeyarwady, and the mapping of major human interventions represent something of a rupture from the preceding narrative on rivers. They are correct. I envisioned spending another two years of intensive field work together with my Burmese collaborators to knit these elements together but was unable to do so, for reasons outlined above. Thus, for understanding the role of *nats* and local explanations for the decline in the fish catch, and especially the hydrology and geomorphology of the Ayeyarwady itself, I have had to rely on the collaboration of Burmese friends and international authorities on the Ayeyarwady watershed. Needless to say, they write with their own voices—often preferable to mine! I hope that the differences in voice are more than compensated for by the knowledge they convey.

ACKNOWLEDGMENTS

I am exceptionally lucky to have three valued collaborators who have played a key role in understanding the river spirits and the local reasons given for the long decline in the fish catch. Together, Naing Tun Lin and Maung Maung Oo have interviewed local fishermen and those knowledgeable about spirit worship along the river; I am forever indebted to both of them for their seminal role in helping me understand the river. Naing Tun Lin is my longtime Burmese tutor and a close friend whose skillful translation and interpretation continue to astonish me. Maung Maung Oo is a knowledgeable guide to life along the river, fishing, and spirit worship. Soe Kyaw Thu contributed some valuable slides that are essential to the book. This study would not have been possible without the joint contribution of these three. Unfortunately, dangers posed by the war have prevented them from traveling much further south of Mandalay; otherwise a more complete inventory of the Ayeyarwady river spirits and fishing experience might have been possible.

Until my retirement in 2021, I was in the habit of offering an undergraduate seminar on rivers every year. The enthusiasm of the students, our close examination of powerful books about rivers, and the research papers the students wrote were important elements in my evolution in the understanding of rivers and their watersheds, river life in general, and fish.

Over nearly a decade I have had a series of graduate and undergraduate research assistants whose curiosity and meticulous analysis enlarged and deepened my understanding of rivers. Michael Lebwohl wrote an essay on the geographical history of what might be called the Paleo-Ayeyarwady. Dana Graef provided me with a discerning survey of virtually all the literature on floods and wetlands. I would never have found my own way to some of the gems she uncovered. Finally, Georg Nieto provided an enormous bibliography of both the Ayeyarwady itself and rivers in general. To top it off, he produced a huge twelve-by-four-foot map of the river that loomed over me while I was drafting this book.

Four geographers, Charles-Robin Gruel, Jean-Paul Bravard, Yanni Gunnell, and Benoit Ivars, have provided quite comprehensive studies of the geomorphology, alluvial islands, pollution, fish populations, and adjacent floodplains of the Ayeyarwady. Their massive reports appear to be unequaled in their scope and detail and are essential sources for anyone writing about the Ayeyarwady. I have poached many of their maps and charts, with their blessings and those of their sponsors.

The reader will note that there are a lot of images, maps, and photos of riverine creatures in the text. I profited mightily in their selection and placement from the help of Pratima Garg and

Jacqueline Ly, both of whom performed acts of technical wizardry of which I was incapable.

The scholars and friends who have taught me what I know about rivers and Burma are legion. Conscious that I have surely overlooked many of them (for which I apologize), I thank those whose work and commentary I have benefited from: Mark Cioc, U Khin Maung Gyi, Soe Kyaw Thu, Hugh Raffles, Martin Doyle, James Prosek, Tharaphi Than, Tun Myint, Anna Tsing, David Biggs, Thabaw Sein, Willem van Schendel, Paul Breton, Ian Baird, Michael AungThwin, Nick Cheesman, Kyaw Hsan Hlaing, David Moe, Ardeth Maung Thawnghmung, Helen Maria Kyed, Mandy Sadan, Eric Harms, and Zali Win.

Finally, I offer my deepest thanks to my long-cherished and exemplary acquisitions editor Jean Thomson Black at Yale University Press, to the exceptionally talented editor/editorial assistant Elizabeth Sylvia, and to the master of images and maps Bill Nelson for helping to bring this book to fruition.

IN PRAISE OF FLOODS

Introduction
A Word about Rivers

Rivers, on a long view, are alive. They are born; they change; they shift their channels; they forge new routes to the sea; they move both gradually and violently; they teem (usually) with life; they may die a quasi-natural death; they are frequently maimed and even murdered. Each river, though subject to the same hydraulic laws, has its own unique personality and history. It makes abundant sense, then, to speak of the life history of a particular river, of its eco-biography. The biography of any river—the Orinoco, the Zambesi, the Mississippi, the Yellow, the Ganges, the Amazon, the Danube, the Ayeyarwady—would be every bit as distinctive as the personal biographies of the various shamans, sages, fishers, philosophers, tyrants, rebels, and saints who lived along their banks. The life expectancy of a river is, of course, typically far, far longer than the lifespan of a single human. This life expectancy has been altered with the invention of dynamite, earth-moving machinery, and reinforced concrete; coupled with the Great Acceleration of the thick Anthropocene, these interventions have drastically increased the mortality and morbidity rate of rivers.[1]

Just as our concept of human biography impels us to think in terms of an individual lifespan, so our sense of the biography of a river should prompt us to widen our temporal lens and think in terms of "river time." Our default unit should be the life of a river. Once adopted, this perspective departs radically from the tendency of *Homo sapiens* to think, at best, in terms of three generations (our parents, ourselves, and our immediate progeny). Other entities—in this case, rivers—operate on a far vaster time scale and can only be understood in these terms. If, instead, we were to focus on the lifespan of other creatures such as fish, insects, or birds, our temporal sense would shrink proportionately, unless we choose to assess the lifespan of an entire species.

The temporal lens we adopt is thus dependent on what it is we want to understand. In geological time—4–5 billion years of it on Earth—all of the rivers in the world are, comparatively speaking, infants. If, as Neil Shubin graphically puts it, we think of scaling down Earth's history to a single year, with January 1 representing the Big Bang and midnight on December 31 the present, then until June there were only single-celled microbes such as algae, bacteria, and amoebae.[2] The first human appears on December 31. A great many rivers, however, appear in their present form only well after the peak of the last glacial maximum ended, some twenty thousand years ago. Until then, so much of the Earth's water was locked up in polar ice caps that the sea level was far lower (roughly 120 meters) than today, and many rivers were a mere trickle of the water they now carry. Later, the huge meltwater pulse over the following six thousand years raised the sea level by roughly 100 meters, making huge bays or lakes and totally submerging many low-lying river deltas, including that of the Ayeyarwady. The subsequent rise of lands, now unburdened by ice, together with the deposition of sediment and a slight decline in sea levels, gradually established the river forms we are familiar with today. Another diagnostic example is the St. Lawrence River in North America. It owes its present form to the rapid melting of glacial Lake Agassiz and the Laurentide ice sheet, coupled with the rebound of a landscape previously compressed by the sheer weight of the glaciers. The result, around ten thousand years ago, was a massive pulse of glacial meltwater eastward along a geological fault line that now defines the channel of the St. Lawrence. This major climatological event is credited by some geologists with an abrupt rise in sea levels of one to three meters, unprecedented floods, and a prolonged cold spell in the North Atlantic. Other examples abound. The point is that, geologically speaking, many rivers are in their infancy. Compared to an individual lifespan, they may seem well-nigh immortal, but they are a good deal younger than our latecomer species as a whole.

What is a river, anyway? For most readers, myself included, the mention of a river's name—for example, the Mississippi or the Nile—evokes the representation on a map of its major channel, its path through the landscape from its headwaters to the sea when it is "well behaved." Looked at through specialist lenses, a river takes on particular attributes of interest. For a hydrological engineer, it may represent so many potential dam sites to generate electricity or suitable sites for levees and spillways to prevent the flooding of valuable real estate. For a public-health specialist, it might represent so much potable water to support the population along its banks. For farmers cultivating the floodplain lands, it might represent vital irrigation water and the deposition of nutritious sediment. For merchants and shipping firms, it might represent a navigable thoroughfare for moving goods upstream and downstream. For a tannery, a cement factory, or a chemical industry, it may merely be a convenient, free sewage-disposal system. All of these perspectives, it should be noted, focus in one way or another on flowing water as a resource to be deployed for the profit or, more charitably, for the benefit of humankind.

The perspective adopted here rejects such anthropocentric tunnel vision in favor of a more capacious understanding of what a river is. First and foremost, we insist on treating rivers as an assemblage of life-forms that depend on the river for their existence and well-being. One, but only one, of those life-forms is *Homo sapiens*. It also includes literally millions of other citizens of the river, from ephemeral insects to waterfowl, mussels, fish, floodplain species of plants and trees, the long-lived river dolphins, and channel catfish. So, if we imagine for a moment summoning a parliament of all the citizens of the river to debate its condition and fate, humankind would be a small minority party.[3] (Chapter 5 attempts to conjure up these other citizens' voices.) Conventional understanding of most rivers is largely confined to what we might call the main stem of the river, downstream of its headwaters and upstream of its delta and distributaries. This common practice seems to derive from mapmakers' inclinations to designate the largest section of flowing water "the river" and then to trace it back upstream until it ends in a glacier, a lake, or a spring, which is then designated the source of the river. The search for the source of the Nile, the Mekong, or the Amazon has claimed almost as many lives of imperial adventurers as has the race to be the first to reach the summit of the world's highest mountains. And yet, the source of a river might just as well be the longest upstream tributary as the one carrying the highest volume of water. Precisely where a river "begins" is arbitrary.

If we consider a river to be an assemblage of life-forms dependent on the flow of water, silt, sand, clay, and gravel—all the elements that we call a river—then our conception of the entity must necessarily include all of its upstream tributaries and all of its delta distributaries. Not only are they all connected as a system of moving water and floodplains, many of the life-forms that depend on the river migrate between the many watercourses and rely on the flood pulse for their nutrition and reproduction. These facts require a systemic view. A river can no more be understood as a trunk or stem than a plant can be understood without considering its leaves and roots and the nutrient flows that unite them.

Biotically speaking, a river in its entirety—tributaries, wetlands, floodplains, backwaters, eddies, periodic marshlands—represents veritable corridors of life-forms. As one moves farther and farther from the river and its floodplains, the concentration of biotic life decreases precipitously. This decrease applies not merely to the fish, bivalves, waterfowl, and turtles that live on or in the water, but to a whole host of riverine birds (for example, raptors and herons), mammals, and reptiles that depend on the biotic richness for their own nutrition. The same holds for the flora: both the flora that live in the water (algae, grasses, and aquatic plants) and the riverine flora that depend on the niche provided by the seasonal advance and retreat of the flood pulse. This floral abundance, in turn, attracts herbivores such as elk and deer and the carnivores (wolves and large cats) who come to prey on them.

If flowing water is the nutritive essence of "riverness," then, perhaps our view is still too narrow. Why, one might well ask, do we overlook micro flows of water that don't make it onto the map because they are too small or seasonal? Every tributary has its own tributaries and these, in turn, the trickles of water that contribute to them. Where one stops mapping is, again, arbitrary. Properly viewed, then, it is the entire *waterscape* that comprises what we conventionally call a river. Such is, or ought to be, our understanding of the term *watershed*. If we think of the watershed as a circulatory system, the micro watercourses from rainfall, from dew, from vernal ponds and springs are the capillaries or, better put, the feeders that undergird the entire system. They provide far more than merely water. Because they span the entire watershed, they gather and transport most of the nutrition that sustains the riverine ecosystem. If one were to exclude their contribution and that of the annual flood pulse, the residual river channel itself would be comparatively sterile biotically.

I find it helpful to visualize the workings of a river as an inverse of the working of a large, leafy plant. Although the plant works against gravity by capillary action, the watershed works entirely by gravity. The roots of the plant, aided by millions of mycorrhizal connections, gather the essential nutrients for plant growth and reproduction. In their function, they are analogous to the millions of sites of wetness in a watershed that slowly aggregate and convey nutrient-laden water to streams and tributaries and then to the stem of the river, its distributaries, and eventually the sea. The foliage and fruits of the plant are, in the same way, analogous to the rich and fertile floodplain (cultivated or not) and the abundant riverine life that is made possible by the concentration of nutrients brought to the river from the entirety of the watershed.

Rivers are "good to think with." For those interested in the Anthropocene and the Great Acceleration, rivers offer a striking example of the consequences of human intervention in trying to control and domesticate a natural process, the complexity and variability of which we barely understand. As George Perkins Marsh understood in his prescient mid-nineteenth-century book *Man and Nature,* rivers are the outstanding example of the earliest human impact on natural systems.[4] The floodplains were virtually everywhere, the productive center of early civilizations. Managing them to promote irrigation, control floods, and transport goods (particularly timber) was central to statecraft. For this reason, rivers were the focus of most early efforts to manage the unruly natural world. From the Roman aqueducts to the Han Dynasty's river masters to the canal bubble of the late seventeenth century in Europe, enormous resources were devoted to the management of rivers for political and economic gain. If, then, we are interested in the history of human intervention into complex natural systems to turn them to human and state purposes, the example of river management offers an ideal metric for the Anthropocene alone.

ONE

Rivers
Time and Motion

E pur si muove.
—GALILEO

Everything Flows
—VASILY GROSSMAN

Everything, literally everything, moves. Nothing, literally nothing, is stationary.

As a matter of daily perception and practice, we safely assume the opposite: that some things move, and other things are stationary. The ground under our feet, the ridge on the horizon, the shade tree that we pass can all be relied upon to stay put. In the case of the shade tree, we know full well that trees, as living things, sprout, grow, mature, and eventually die, but on a day-to-day or week-to-week time scale, the changes are largely imperceptible. A tree is, for the time being, a steady part of the landscape.

Apprehending the universality of movement is almost entirely a matter of widening the temporal lens of our perception to encompass massive movements that are invisible on a shorter time scale. The widest opening imaginable, I suppose, is "galactic time," in which the life (and impending death) of our own miniscule solar system is but a minor episode (starting a relatively recent 4.6 billion years ago) in the 13.5-billion-year history of the Milky Way galaxy. At a more modest, though scarcely more comprehensible, "earthly" time scale, we are concerned with geological time. Research into the movement of tectonic plates has, since the breaking apart of the supercontinent of Pangaea roughly 335 million years ago, traced the massive movements that created the global map of continents that we now take for granted. However imperceptible to us, we are being moved slowly, a couple of centimeters or so a year across the globe, lifted to a higher altitude or lowered closer to sea level. The Atlantic Ocean is widening, pushing the Eurasian and African tectonic plates further away from the North Atlantic plate.

The most convenient method for capturing movement over vast periods of time is, as shown in the accompanying figures, through a temporal series of snapshots, ordered chronologically, as in time-lapse photography. It should be noted that the effect of movement is achieved by juxtaposing *still* representations from successive moments. The assumption that the viewer is likely to make, unless forewarned, is that the pace of change between each of these snapshots is steady and uniform. This assumption may, in fact, be false.

Reducing the scale further, the same process is employed in representing glacial and interglacial periods in Earth's history. Here, the opening of the temporal lens to hundreds of millennia allows us to grasp the massive advances and subsequent retreats of the polar ice caps and the climatic epochs they represent. One such representation depicts shifts in glaciation from 123,000 years ago until nearly the present.

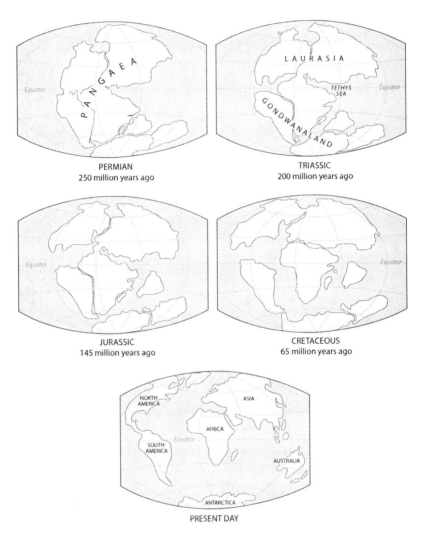

Movement of tectonic plates from the supercontinent Pangaea 250 million years ago to their position today.

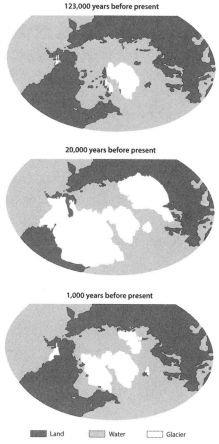

Advances and retreats of glaciers over North America and northern Europe from 123,000 years ago until 1,000 years before the present.

Moving from galactic time to tectonic time and then to glacial time is, in one sense, a radical narrowing of our temporal scope. Glacial time at least enters the realm of what we might call *Homo sapiens* time, inasmuch as anatomically modern humans are now believed to have appeared not much earlier than two hundred thousand years ago. Although this conception of *Homo sapiens* time is compatible with recent archaeological findings, most of our fellow species would be more inclined, along with historians, to start the clock on *Homo sapiens* time roughly when the earliest cities in the Fertile Crescent began to leave evidence of monumental architecture, luxurious grave goods, and the first evidence of writing, no earlier than twelve thousand years ago.

Even at this cosmically minute time scale, the temporal lens is still wide enough to encompass the movement of things like shade trees that we regard as fixed and stationary. Oak and beech trees, in the grips of the cold and dry conditions of the last glacial maximum, roughly 20,000 years ago, were confined to small refugia in the mountains of the Iberian Peninsula, Italy, and the Balkans.[1] As the climate grew warmer and more humid, around 14,500 years ago, they began to migrate northward by the dispersal of seeds and pollen and, with a brief setback due to the cold spell of the Younger Dryas (12,000–11,600 years ago), they continued to march north apace until at least 8,000 years ago. Given the time, a favorable climate, pollinators, and birds, they were as mobile as the first *Homo sapiens* who crossed the Bering Strait to the Western Hemisphere. A time-lapse representation would show the uneven but cumulative march of these floral settlers colonizing the northern latitudes. And, like many colonists, they were also shapers of a new landscape. They moved collectively with a whole suite of fellow travelers: soils, microorganisms, insects, pollinators, birds, and fauna, including *Homo sapiens*. Deciduous forests of oak and beech replaced more cold-tolerant conifers that in turn also colonized higher latitudes previously dominated by steppe and tundra ecologies.

These brief excursions into radically different time frames serve to highlight a world of movement and change that is nor-

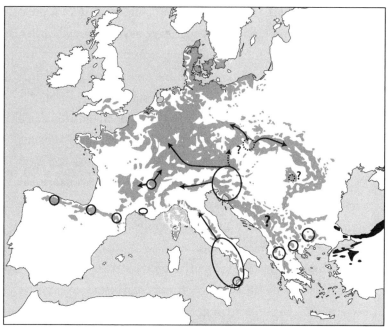

Proposed refuge areas for European beech (*Fagus sylvatica*) during the last glaciation and routes of recolonization after the glacier retreated. The circles indicate the main source areas, and the arrows show expansion after the retreat. The question marks and dotted lines denote hypothetical or speculative refuges and postglacial spread.

mally ignored in our quotidian lives. They also demonstrate the subversive power of history to teach us that what we take as given was not always the case and might not, therefore, always remain the case. The deeper the history, the broader the time frame, the more of our world is in motion—and the more of our certainties are called into question.[2]

The default time frame or temporal optic for humans is, as one might suppose, a single human lifetime. A human life is the unit that defines us. Autobiographies, memoirs, and biographies operate self-consciously within these confines, as does a great deal of fiction and film. More generously, as I mentioned earlier, we might set our species default temporal optic at three generations. For an account of individuals and families, such an optic is entirely appropriate. For an account of "our kind," however, it is far too, well, myopic. The lenses appropriate for an individual of a species and the species as a whole diverge radically. Elephants and parrots may live long lives as individuals compared, say, to dragonflies and starfish, but their collective existence as a species is much shorter. The same is true for the world of flora; an individual oak tree may live a far longer life, but as a species, it may be far younger than a horsetail fern.

If we look beyond the span of an individual life to the career of a species, as naturalists and environmental historians urge, we are certain to encounter changes and movement that are not observable over a single lifetime. But what if the object of our attention is not a particular species but an assemblage of species and their habitats? Such would be the case if we were to study a mountain, a forest, a city, or a polar ice cap with all the species for which that habitat is their lifeworld. This is precisely the objective we aim to pursue with respect to rivers: their history, their movement, the species to which they are host, the way in which they have evolved or been altered by human intervention, and the consequences for the assemblage as a whole.

The River as Movement

Mark Twain in *Life on the Mississippi* commented,

> One who knows the Mississippi will promptly aver—not aloud, but to himself—that ten thousand River Commissions, with

the mines of the world at their back, cannot tame that lawless stream, cannot curb it or confine it, cannot say to it, Go here, or Go there, and make it obey; cannot save a shore which it has sentenced; cannot bar its path with an obstruction which it will not tear down, dance over, and laugh at. But a discreet man will not put these things into spoken words; for the West Point engineers have not their superiors anywhere; they know all that can be known of their abstruse science; and so, since they conceive that they can fetter and handcuff that river and boss him, it is but wisdom for the unscientific man to keep still, lie low, and wait till they do it.

The insistence on movement, on fluidity, as the very essence of rivers would not require such emphasis were it not for the fact that our visual and cartographic grasp of the world implies stasis. We apprehend the river, for the most part, at a distance, as a definitive line on a map, as a snapshot or landscape painting of a river peacefully gliding along its customary banks. A quite false sense of constancy and stability is likely to ensue. That line on a map implies, as Dilip da Cunha so convincingly argues, a sharp demarcation between land and water.[3] In the case of the Ganges, that demarcation is obliterated during the monsoon when the Ganges spreads out over its floodplain. Such is the case for most rivers that are fed by more or less predictable snowmelt or seasonal and monsoon downpours. Huge rivers such as the Amazon at flood season in late April are routinely forty times wider and twelve to fifteen meters above their October lows. The line on the map drawn to represent the course of the Amazon can scarcely do justice to this enormous fluctuation. Those who live along its banks are only too aware that the Amazon does not respect that line on the map!

As with most natural phenomena, if we open the temporal lens to its widest aperture, it becomes clear that the rivers we identify today, such as the Zambesi, the Ohio, or the Mekong, were, if they existed at all, quite different in earlier millennia. Their course may have been radically changed by earthquakes or volcanoes or glacial melt or by the slower but relentless movement of tectonic plates that diverted the channel in an entirely new direction. The Ayeyarwady River of Burma, which we shall examine in more detail later, was diverted by the volcanic eruption of Mount Popa in recent geological time and then, later, shifted westward from its previous course along what is now known as the Sittaung River to a new course to the sea. And just to further confound our sense of the Ayeyarwady being a line on a map, in the case of a major flood, the Ayeyarwady spills over to reoccupy its ancient course along the Sittaung.

Climate changes in geological time, and most particularly the Holocene warming effect on sea levels, inundated many low-lying rivers, traces of which can still be detected on the sea floor. Rivers that had not even existed at the height of the last glaciation became raging torrents. A glacier is, after all, a slow-moving river of ice. Long before our current preoccupation with human-caused global warming, the natural movements of the Earth and its varying angle to the sun created, we now understand, long cycles of warming and cooling that gave birth to rivers and then froze them in place.

The movement of rivers long antedates the recent appearance of *Homo sapiens* as would-be shapers of rivers. We have thus far noted what might be called the external causes of river movement. What we have ignored are the myriad ways in which a river, as a hydrological process, produces movement by its own action. Looked at in terms of downhill flow, a river is a huge transmission belt and receptacle, moving silt, sand, clay, gravel, and plant matter from higher elevations to lower ground. The steeper the pitch, the faster the flow and the greater the amount of material it delivers to lower elevations. In doing so, it carves out new channels, creates new floodplains, and deposits enough soil-building material to extend and reshape the coastline, thus even further obliterating the sharp dividing line between land and water on the map.

Left to its own devices, with no human interference, it is natural for a river to block its own route to the sea and then "jump" to a new bed. This happens most commonly when a river descends to a flatter coastal plain. As the speed of the current slows, the material it has scoured from upstream is deposited on the riverbed, raising its elevation. The more sediment the river carries and the lower the gradient of the coastal plain, the faster the bed of the river rises. A river thus creates its own levees and barriers; the coarser materials—gravel, pebbles—are deposited first, the less coarse material—sand and finer gravel—are deposited further along, and the finest silt and clay are likely to settle out last. Eventually, the elevation of the riverbed slows the current to the extent that it blocks its own way forward. The water is diverted laterally as it seeks lower terrain to speed its way to the sea. In turn, the new channel's dynamics are likely to repeat the process, such that a still newer passage to the sea is carved.[4]

Nothing illustrates this process better than the millennia-long vicissitudes of the Yellow River, the cradle of the earliest Chinese kingdoms, on the flat North China coastal plain.[5] The channel swung north and south of the Shandong Peninsula as it repeatedly blocked its own way to the sea. Its very name is a reference to the massive load of sediment it carries and deposits.

The accompanying figure depicts the many courses followed by the Yellow River over the past three thousand years. In the last two and a half millennia, it has shifted no fewer than twenty-six times, swinging north of the Shandong Peninsula to the Bo Hai Bay and then south to enter the Yellow Sea south of the peninsula, and then back again. The distance between the resulting estuaries was as great as eight hundred kilometers.

Another figure traces the many course changes over the less than three centuries between 1048 CE and 1321 CE. Once again, the time-lapse snapshot quality of this delineation of dry land from the river obscures both the erratic convulsions of floodwaters and the consequent swelling and shrinking of wetlands.

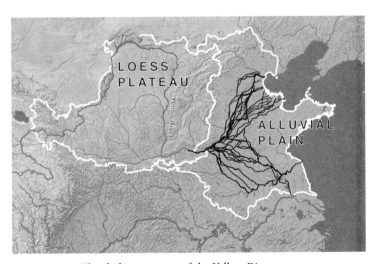

The shifting courses of the Yellow River over the last three thousand years.

The key factors behind the apparent capriciousness of the Yellow River are the combination of sediment, water volume, and, above all, the extremely low gradient of its floodplain. Virtually all the significant course changes originate from a single extremely flat region around Kaifeng, located at the western reaches of the floodplain. The very flatness of the region makes it especially prone to floods; even a small rise in floodwaters will, facing no barriers, inundate wide areas of the plain. Add the routine "natural" annual load of sediment from the highly erosive loess soils upstream, and the path of the river becomes even more erratic. When, on top of this process, one adds the massive deforestation for military and agricultural settlement purposes documented by Ruth Mostern, the sediment brought to the plain rises exponentially and the river writhes back and forth as channel after channel is plugged.

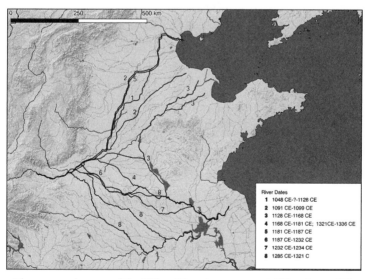

The course changes of the Yellow River between 1048 CE and 1336 CE.

As the bed of the river near Kaifeng rose with sediment, so did the dikes and levees designed to protect the adjacent (taxable and conscriptable!) agricultural populations. The result was what one historian has called "technological lock-in," in which the enclosure of the river causes a corresponding rise of the riverbed and, in response, a further raising of the protective levees.[6] The paradoxical result is that the bed of the river is raised *above* the surrounding floodplain. In the case of the Yellow River at Kaifeng in Henan Province, the bed of the river is more than ten meters above the plain. What is one to make of a situation that so defies our traditional understanding of a river? Common sense tells us that what we have here is not so much a river as an *aqueduct*.[7] Extreme as this example is, the process of sediment deposits raising the bed of a river and the natural formation of levees by annual flooding is, at bottom, a natural process predating the appearance of *Homo sapiens* on the scene. What is distinctive about the Yellow River's history is the additional and massive erosion of friable loess soils by agricultural and military activity and the reliance upon higher and higher levees in an attempt to shield the taxpaying subjects from crop-destroying floods.

Reflection on River Time

Harold Fisk, a geologist working for the U.S. Army Corps of Engineers, published a remarkable series of maps in 1944 tracing the constantly shifting paths of the Mississippi River from prehistoric times to the mid-twentieth century. His reconstruction of the river's convoluted history from southern Illinois to Louisiana was based on some sixteen thousand boreholes, some of them as deep as thirteen thousand feet (nearly four thousand meters), to recover the successive layers of soil and sediment. The resulting color-coded maps (which we have had to reproduce here in black-and-white) are so visually arresting that they are routinely sold as objects of display. Even in black-and-white, however, they represent a rare and mesmerizing representation of river movement, of "river time," across more than three millennia. Reproduced here are maps number 6 and 7 of the Mississippi "meander belt," covering the river's ancient courses from Marion, Arkansas (opposite Memphis) downstream to the floodplain near the delta town of Greenville, Mississippi.

Viewed through this lens of "river time," the maplike delineation of water from land is untenable. The Mississippi has, over time, refashioned itself thousands of times and, in so doing, refashioned the surrounding landscape for all riverine creatures, including *Homo sapiens*.

What Fisk's geological research and maps capture so well are the major—one might even say epochal—changes in river morphology. This focus on the *longue durée* necessarily has to ignore the day-to-day shifts in shoals, sandbanks, and river channels brought about by smaller disturbances. In the course of my research travels in Burma, I had occasion to experience the effects and practical importance of such seemingly micro disturbances. The Ayeyarwady River, which virtually defines the alluvial heartland of Burman culture, is at its most shallow between December and March before the onset of the monsoon rains. Navigation in this period is particularly hazardous, and the larger river passenger and cargo ships are prudent about limiting their draft, lest they run aground. This caution is particularly vital when vessels are traveling downstream for two reasons. First, they must exceed the speed of the current or else lose the ability to steer. Second,

Mississippi River meander belt (map 6).

should they happen to run aground in a sandbank, the pushing action of the current further embeds them in the bank, impeding their ability to back off and free themselves.

One might imagine that, equipped with a detailed map of the riverbed, an experienced captain would be able to navigate it suc-

Mississippi River meander belt (map 7).

cessfully. This is, alas, not the case, inasmuch as the riverbed is changing constantly and unpredictably. In the course of two trips on a small inboard motorboat carrying roughly twenty passengers from Mandalay downstream to the ancient ruins of Bagan, I had occasion to appreciate the dangers and the steps taken to avoid them. The captain had plied these waters for at least a decade, but he did not trust himself to navigate without expert help. Over the eight-hour trip we took on four different pilots, each of whom had a unique and intimate knowledge of a particular stretch of the river channel. When we reached the limit of one pilot's local knowledge, he debarked by arrangement at a village where a second pilot, who knew the next stretch of the

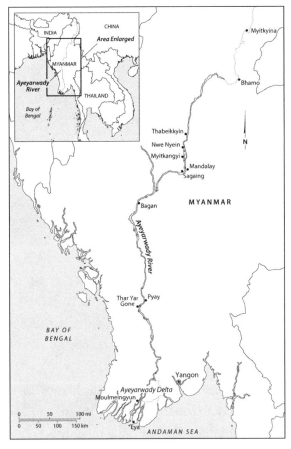

Map of the Ayeyarwady River.

river ahead, came aboard. He was, in turn, replaced after roughly two hours by a third and then a fourth pilot until we arrived at Bagan. The changes in channel conformation were so frequent that the captain needed a "local tracker" for each stage of the trip to guarantee safe passage. Even then, he ran aground once or twice each dry season.

During the low-water period, larger passenger and cargo ships either forgo sailing altogether or drastically reduce their cargo to minimize their draft. It is not unusual, even then, for pilot error to result in a grounding. In the most severe cases, when the ship cannot be pulled free by tugs, it simply remains aground until the next monsoon rains are sufficient to float it free. In many cases, such ships, although they are carrying a cargo of valuable teak logs (which barely float), are simply abandoned for the season. In the case of other cargo, it is frequently off-loaded to smaller craft, and a skeleton crew is left aboard to prevent the ship from being stripped by scavengers of valuable machinery, electronic gear, and pipes.

On one occasion, I was on a larger passenger vessel traveling downstream that ran aground. After several hours of reversing engines and trying to wriggle free, three crew members were put off to port, perhaps thirty meters from the stern of the ship. There, in very shallow water, they embedded a large stake into the riverbed and attached a cable to it, in turn attaching that to a motor-driven winch at the stern of the vessel. Once in place, the apparatus allowed the captain to wriggle the stern to portside and, after an hour of effort, dislodge the bow and back away from the sand shoal. This was, I learned, a fairly standard maneuver, emphasizing the residual unpredictability of the riverbed.[8]

Three men working to free a grounded vessel in the Ayeyarwady River.

Constantly shifting channels and underwater shoals are a distinctive dry-season navigation problem. Sometimes in this season, larger upstream-bound vessels cannot proceed, as in the case of the Ayeyarwady near Bhamo (Bamaw); the ship must anchor and transfer the cargo to smaller craft that can navigate the shallower water. These concerns evaporate during the monsoon high-water months (August, September), when the volume of water is roughly eight times greater than its dry-season level. The river is then deep enough that pilots are no longer needed. Not only is the current-aided trip downstream much faster and much safer, the trip upstream is also faster, despite having to contend against a much swifter opposing current, simply because there is no longer any need to follow the twists and turns of what, in the dry season, is often the only navigable channel.

The River Makes Its Path by Flowing

Theodor Schwenk, in *Sensitive Chaos,* observed,

> Water flows and streams on the earth as ceaselessly as the stream of time itself. It is the fundamental melody. . . . Unremittingly it belabors the solid earth, grinding, milling, destroying, levelling out and at the same time building up again, creating anew. . . . As the life blood of the earth, in the great network of veins, it shifts unbelievable amounts of substance, which everywhere accompany the life processes of the earth and its creatures. . . . It transforms the hardest rock and the highest mountains, and it dissolves finished forms, preparing for new creation. Water is the great exchanger and transformer of substances in all forms of metabolism.[9]

To see the flowing water of rivers as a central factor in this transformation of landscape requires precisely the very wide-angle temporal lens of river time that we have urged. It takes a long time for flowing water to disassemble a rock! And if the process is attributed solely to water—that is, to H_2O alone—it overlooks the vital role of life-forms, especially bacteria, which are integral to the "milling" effect of water.

Once again, the term *watershed* is far more evocative of this process than the term *river,* if for no other reason than "river" has come to connote whatever is designated as the main channel of a

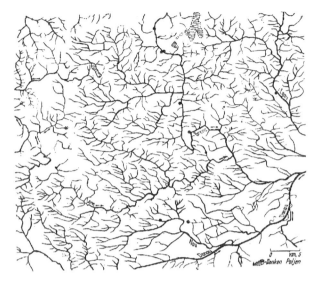

Map of water veins in the catchment basin of the Brenz River, a tributary of the Danube in Germany.

named waterway on a map. This narrow understanding of a river omits its tributaries, the tributaries of *these* tributaries, and so on, even to the tiniest seasonal rivulets that feed these tributaries. Thus, a complete map of a river would require a minute filigree that would trace each and every discernible watercourse. One effort to do so for the Brenz River, a tributary of the upper Danube, gives a sense of the minute detail that would be required.

Even this degree of reticulation, as I have argued, is still not sufficiently comprehensive. As I suggested earlier, taking the "water" in *watershed* literally would require thinking of a waterscape covering the watershed in which, say, wetlands, damp soils, tiny puddles, ice, dew, and groundwater in general would all factor in. As forms of wetness, they are all connected elements of a single hydrological system.[10] This connectivity afforded by water will be central to much of our reasoning about the life-forms whose habitat is the river. For the moment, however, it is enough to observe that in most watersheds, connectivity fluctuates seasonally from a maximum at high water to a drier season when, due to evaporation, the degree of connectivity is much reduced.

Meanders as Movement

The term *meander* comes to us from an actual river in west central Turkey, the Büyük Menderes, which follows a winding course over a flat plain before entering the Aegean. The Menderes appears in Homer's *Iliad*. As a verb in English, the word has come to mean aimless wandering in walking, speaking, or writing. As a technical geological and hydrological term, however, it denotes a distinct nonrandom and rather systematic pattern of movement. A meander might rightly be described as a species of wave and, like a wave, it can be described in quantitative terms. The formula for doing so is a simple calculation of the degree its form diverges from a straight line. Thus, when a sinuous river is mapped, its course is compared to a straight line that fits its trajectory.

The straight-line river would be described as "1." If the actual river, measured by the length of its actual channel, is as much as 50 percent longer (1.5), it is classified as sinuous. If it is any longer—say, 2.3 times as long—that ratio is the index of its "meanderness."

How meandering a river is depends on many factors, such as the width of the valley through which it passes and the erodibility of the land it crosses. All else being equal, the more gradual the gradient, the more pronounced the meandering. If we could

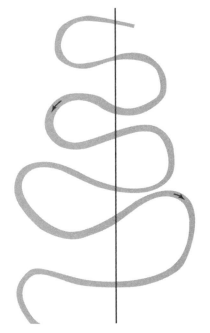

Measure of Meanderness = ratio of river channel length to straight line

Meanderness of a river.

imagine a perfectly smooth, absolutely flat, and evenly tilted surface, water would run off it in a uniform sheet. But no such surface exists in nature. The irregularities of the surface and gradient always initiate a process of uneven flow—and hence of meandering.

The meandering river is a radically different river than its straight-line point of comparison. The die-straight river has a steeper gradient and a higher water velocity and is capable of carrying large loads of sediment great distances. Meandering, by contrast, creates a river that is longer and slower; it deposits a great deal of sediment that invades the floodplain to create adjacent wetlands and extensive habitats, as we shall see.

What deserves special emphasis in this process is natural movement over time. Rivers have typically been meandering since long before *Homo sapiens* appeared on the scene.[11] It is not simply that the water is moving and carving its way forward: in the process, it is remaking the landscape as well.

Once even a small curve forms in a river or stream, the process of intensification of a meander is initiated. The flow of the water along the outside concave bank is faster and therefore more erosive as it carves away at the bank. In doing so it sweeps more sediment from the riverbed and creates deeper pools. Water flowing along the convex side of the meander loop is, by contrast, moving more slowly and therefore depositing more sediment, creating what is called a point bar. The entire process is self-intensifying. As the external bank swallows more of the terrain, its velocity increases, while the inside point bar velocity diminishes, thereby speeding up the process of sculpting an ever more exaggerated loop. The process is not a steady, even, day-to-day progression. On the contrary, much of the meander formation occurs when the river is at or near flood stage and when the velocity of the flow and volume of sediment are highest. While meander formation is a continuous process, it advances in practice by leaps and bounds (avulsion) rather than by accretion. In the same fashion, the standard representation of relatively uniform meander loops, as depicted in the accompanying figure, is based on a uniformity of terrain and slope that is rarely the case. Most rivers, in the course of traversing an apparently featureless plain, encounter different soil structures and rock formations that offer variable resistance to the uniform sculpting shown in the illustration. This irregularity is quite apart from massive avulsive events,

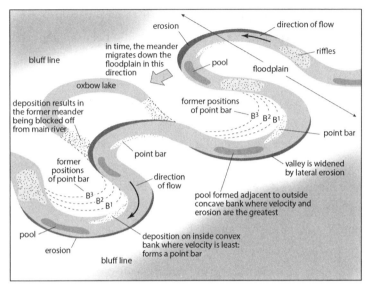

Diagram of a river meander.

such as earthquakes, that totally transform or destroy a riverscape and watershed.[12]

The detached oxbow provides an apt illustration of the combination of constant patterned movement on the one hand and irregularity on the other. The static impression of a river on a map occludes both of these essential features. The hydrological forces of erosion and sediment deposition operate to exaggerate the meander loop so that the "collar" of the oxbow, all else being equal, becomes narrower and narrower. Eventually, the water at flood stage typically breaches the narrow isthmus of the collar and pours directly into the downstream river, carving a new route with its relative velocity and volume, isolating the meander loop, which has now become an "oxbow lake." But perhaps not permanently! At flood stage the next year, part of the flow may well return to the oxbow loop, temporarily reincorporating

Meanders of Nowitna River, Alaska.

it as a separate channel. The oxbow has become a periodically recharged wetland and part-time river channel. And if, for some reason, the sediment buildup on the river channels adjacent to the oxbow raises it a bit higher than the older oxbow channel, the river may reclaim it as its main channel once again. The many abandoned channels depicted in Fisk's color-coded maps of the Mississippi meander belt are quite likely to have become the river channel more than once in the river's long history. This helps remind us again that the entire complex of meanders is itself a vast conveyor belt moving slowly downstream.[13]

The Shapeshifting River versus the Hundred-Year Flood

The emphasis on unceasing movement and the shapeshifting of rivers is intended to amplify the most expansive interpretation of Heraclitus's dictum: "You cannot step twice into the same rivers; for freshwaters are forever flowing in upon you." And yet, implicit assumptions of stasis, regularity, and predictability nevertheless abound. Nowhere is this more evident than in the concepts of the fifty-year flood, the hundred-year flood, and even the five-hundred-year flood that are routinely used in the popular press to describe particular floods. It was a cause for wonderment when the Rhine experienced no fewer than four hundred-year floods in the space of a dozen years: 1983, 1988, 1993, and 1994. Strictly speaking, this was not statistically impossible. The concept of the hundred-year flood does not mean that such a flood will occur only once in a century. It means instead that every year there is roughly a 1 percent chance that such a flood will occur. Thus, over the course of thirty-three years, there is a one in three chance of such a flood. An individual toss of a coin has a fifty-fifty chance of being heads or tails, but three consecutive heads do not reduce the odds that the next toss will not also turn up heads.

The larger question is the statistical base from which the probabilities of 1 percent of such a flood occurring are calculated. If one has reliable flood data going back, say, two hundred years, the calculated flood risk will be much less reliable than if the flood data spanned five hundred years, even under the (unwarranted) assumption that the river hydrology remains constant. To compute the risk of a five-hundred-year flood with a high degree of reliability would require, in principle, perhaps a millennia of flood data. With the possible exceptions of the Nile, Euphrates, Yellow, Po, and Yangzi Rivers, no such longitudinal data are available. For most rivers, not to mention their many tributaries, the existing time series of water levels simply do not permit reliable calculations of this kind.

The truly fatal flaw of most (not all) estimates of flooding is that they assume that they are applying their calculations to the same river, to the same hydrological system.[14] Nothing could be further from the case. As we have repeatedly stressed, the river is always moving, and owing to the thousands of variables at play, it will always defy predictions. The caution against assuming that the present course of a river is an adequate description of its future, visually illustrated in Fisk's striking maps of the Mississippi's historical courses, applies. George Everest, who led a geographical survey of India in the mid-1800s, wrote this about maps of the Ganges: "No map can delineate [the Ganges's] true features for any length of time.... Whenever survey of such wandering rivers be undertaken for purposes purely geographical, the chief object should be to select certain limits within which the river has ever been known to flow, and which there is no chance of its ever passing—to lay down those limits with accuracy—and to treat the existing course of the river as quite a subordinate affair."[15]

Two things are noteworthy about Everest's wise caution about a definitive mapping of the course of the Ganges. The first is that even specifying the "limits within which the river has ever been known to flow" requires an extensive time-series data base that would have been unavailable in the mid-nineteenth century. The second is that Everest appears to be referring to natural (that is to say nonanthropogenic) movements of the river. It is, in fact, virtually certain that the hydrological character of the river had already been transformed by human settlement, deforestation, small-scale drainage, and irrigation works. Decade by decade, century by century, human activity had been altering the river, adding a new and volatile layer to its hydrology. The

Anthropocene for rivers began long before the Great Acceleration contribution of dynamite, earth-moving machinery, and reinforced concrete totally transformed rivers.

Perhaps the major reason why, for much of the last two millennia, you cannot step into the same river twice is that humankind has radically changed the ecology of rivers and their surroundings. Perhaps the most important intervention was the deforestation of the floodplain by agricultural settlement, pastoralism, and the consumption of timber for warmth, cooking, building, metallurgy, mining, and kilns to fire pottery. Of all the ecological systems disrupted during what might be called the thin (or slow) Anthropocene before the late nineteenth century, rivers were the most affected. By 1900, along the Danube River, 95 percent of the upper floodplain had already been lost, roughly 75 percent of the floodplain in the middle and lower Danube, and 30 percent of its delta floodplain.[16] The destruction of the floodplain, of its moisture-absorbing and moisture-holding wetlands and forests, transformed the Danube into a radically different river than, say, the Danube of 1750. It carried far more sediment; the runoff from rain and snowmelt was far more rapid and, of course, it was far more prone to large-scale flooding. Not only did *Homo sapiens* try to domesticate and landscape rivers for their own purposes, they also displaced the other mammalian landscapers whose effects on everyday ecology were much more benign: beavers. The rapid extermination of beavers in North America and northern Europe for the world fur trade had massive effects on the hydrology of rivers, not to mention the destruction of a beaver-created habitat that promoted biodiversity. This human-caused change in the shape and flow of rivers, already subject to constant and sometimes radical change, added a new level of radical unpredictability.

Exposed to the vagaries of rivers, human agents have, usually in vain, attempted to calculate the degree of variation, to put confidence intervals around the uncertainty. Rather like Lloyds of London betting on a thousand sailings, confident that only, say, less than 2 percent will end in disaster, the Army Corps of Engineers, relief agencies, and insurance companies adjust their policies in recognition of a given sharply defined level of uncertainty. As we have seen with the concept of fifty- or hundred-year floods, the data on which such calculations rest presume a level of continuity and stasis that has proven untenable. As the pace of landscape modification has grown exponentially along with population growth, deforestation, and mining since the beginning of the industrial revolution, these quasi-scientific estimates have proven increasingly unreliable. And finally, the runaway pace of carbon dioxide emissions has added a potentially catastrophic layer of uncertainty about rainfall, storms, drought, and evaporation that affect a river's behavior. Unknowable thresholds of uncontrollable effects make for a future of hyper-radical uncertainty.

TWO

In Praise of Floods
Moving with the River

No mud, no lotus.
—BUDDHIST SAYING

No flood, no river.
—JAMES C. SCOTT

The Flood Pulse

The annual flood pulse is the most consequential movement of a river for all the life-forms dwelling in and around it. Whether impelled by monsoon rains, snow or glacial melt, or seasonal rains, the flood pulse represents that part of the year during which the river overflows its channel banks and inundates the adjacent floodplain. It may, year by year, vary in its amplitude, its timing, and its duration. But it is a completely natural part of the annual cycle of any river's hydrology that has not been disrupted by human intervention.

The periodic inundation of the floodplain represents the lungs of the river in an almost literal sense. That is, the flood pulse is the precondition of its vitality and of all the aquatic and riverine forms of life that depend upon it. Without the annual occupation of the floodplain, the channel—that line on the map—is comparatively dead biotically. "Flood" as a scare word is so deeply anthropocentric that I would be inclined always to qualify its use. Notwithstanding that flooding is, for humans, the most damaging of "natural" disasters worldwide, from a long-run hydrological perspective, it is just the river breathing deeply, as it must. On this view, we would regard the flooding of human settlements on the river's floodplain as the result of *Homo sapiens* encroaching on the natural territory of the river—an act of "trespass."

The periodic inundation of the floodplain is, in sum, the lifeworld and condition of existence of all the species that inhabit the river and dwell in its riparian zones. Fish are both the most obvious and most studied example of this dependence on the flood pulse. Many fish that disperse to the floodplain during the pulse get as much as 80 percent of their total annual nutrition in this brief period. They gorge sumptuously on the invertebrates, decaying organic matter, and microbial life, putting on weight and spawning. Huge migrations of fish, both freshwater and seagoing, rush to take advantage of this feeding frenzy. That the floodplain may be as much as forty times the width of the channel conveys some sense of the range and volume of nutrition it places at the disposal of fish no longer confined to what the mere channel has to offer. Quantitative studies document the generative effects of the flood pulse. The fish catch in the Mississippi basin had declined 83 percent over the previous half century until the year after the great flood of 1993, when the catch set a new record. Long-term studies of the Danube have shown that the greater the extent of the flooding in any given year, the greater the haul of fish the year after. This quantification of commercial fish yield, termed "the flood pulse advantage," has been demonstrated both for tropical and temperate rivers.[1]

The flood pulse does not attract only fish. Its nutritional magnetism draws in an entire cavalcade of creatures and flora: waterfowl, riverine wetland birds (for example, herons), migratory birds, muskrats, wolves, raptors, and herbivores coming for the fresh grass that sprouts as the food recedes. And for all the creatures that cannot relocate to the floodplain, as the pulse drains back into the main channel, it brings with it much of the nutrition that the channel receives annually. For the relatively stationary freshwater mussels and clams and the channel catfish, often too large to leave the channel, the receding water delivers a burst of microbial and decaying matter that helps sustain them. As noted earlier, without the runoff from tributaries, rainfall, and the draining of floodwaters back into the channel, the river itself—that line on the map—could sustain but little life. The river, in this sense, is fed by the entirety of its watershed and, especially, by periodic floods.

The term *flood pulse* was coined in 1989 by Wolfgang Junk and his colleagues.[2] Because the concept muddies (pun intended!) the traditional distinction between terrestrial and aquatic ecosystems, it is paradigm-shifting. It defines and examines a *vast in-between landscape* that is transitional, periodically inundated, periodically dry, and periodically damp. In this transitional zone are found flora, insects, birds, and mammals adapted to its periodicity and fluctuations. Virtually endless variations are to be found within this pulse zone: areas that are inundated for much of the year, areas that are inundated only in exceptionally high-water years, and every imaginable intermediate possibility. The assemblage of life-forms adapted to the pulse zone will also depend on temperature, soil structure, and terrain. What is clear is that a river channel cannot be understood without its periodic lateral movement. That line on a static map once again fails utterly to capture the essence of a river's life in movement. Only a long series of time-elapse aerial photographs can begin to capture this vital lateral movement.

For professional ecologists, the flood pulse represents a key example of what they term disturbance ecology. Such disturbances—flood and fire are iconic in this context—create a patchy mosaic in which a new succession of colonizing plants and creatures adapted to pioneering areas temporarily cleared of competitors can establish themselves. A rich mosaic of patches at different stages of succession provides a landscape diversity that an absence of disturbance would preclude. Disturbance, then, makes a major contribution to landscape diversity, which, in turn, affords the life-forms within this mosaic the benefit of resources, safety, and nutrition of adjacent patches. In this sense, one could also speak of a "fire pulse." The key difference is that the flood pulse on an unengineered river is more predictable. This means that, like fish that dwell in the tidal zone, the same patch of floodplain, depending on the season, offers a wide range of environments as the floodwaters rise and then recede, leaving ephemeral vernal wetlands. This helps explain why a watercourse and its floodplain, as a system, represent the highest biodiversity of any freshwater ecology.[3]

Floodwaters, as they move across the landscape, create a huge variety of habitats: backwaters, ponds, marshes, swamps, slow-moving warm water, refuges from predators, and assemblages of food and habitat that favor a large variety of riverine species. At bottom, it is a story of habitat and nutrition. The entire

Examples of lateral shift of the Ayeyarwady River during
the monsoon: (*left*) May 28, 2007, and *(right)*
September 26, 2007.

mechanism depends on the microbial richness of the floodplain, which represents the base of the food pyramid in the lifeworld of the river. The phases of the flood pulse have been condensed and illustrated in the accompanying figure.

In some ways, especially to lay readers, the term *disturbance* carries misleading connotations. A disturbance is typically understood as an interruption of a settled and peaceful condition, as in "disturbing the peace," where peace stands for a legal description of normal order. But for ecologists, the disturbance caused by annual flooding and natural fires is seen as normal and largely beneficial. It bears constant emphasis that the periodic inundation of the floodplain by a river at high

water is part of its natural movement. Indeed, as we shall see, to artificially prevent flooding by use of levees, dikes, or dams is to create a *true* disturbance by preventing a river's natural movement.

The movement of water across the floodplain landscape is the engine of a river's life-giving properties. It stimulates the rapid growth of marsh and swamp plants, algae, and aquatic plants (for example, the lotus), and this, in turn, drives a spurt in the growth of herbivores and insects (mostly tiny crustaceans, zooplankton, and aquatic insects), providing a feast for spawning fish and their newly hatched fry.[4] As the flood, which in some cases may occur more than once a year, recedes, it typically exposes fresh, nutrient-rich soil ideal for the growth of semi-aquatic and terrestrial annual grasses. This provides another fresh feast, this

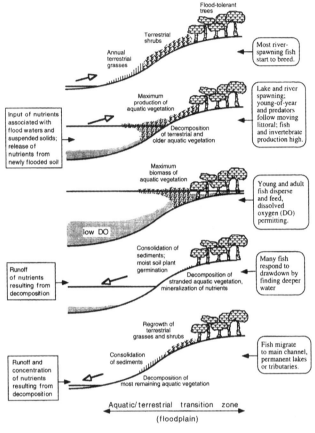

The flood pulse.

time for herbivorous fauna and birds, while the retreating water pours its periodic dose of nutrition back to the channel-dwellers. As Seth Reice summarizes, "Flooding therefore enhances both terrestrial and aquatic productivity. The cycle of flooding is essential to the life of every river."[5] This explains why rivers and their floodplains, compared to purely terrestrial settings, constituted the green corridors of life and movement long before the interventions of *Homo sapiens*.

The key to the biodiversity and productivity of the river-floodplain system is what ecologists describe as connectivity. In simplistic terms, it is the extent of the terrain that is watered by a flood. More technically, surface connectivity refers to how much flow the floodplain and the secondary channels receive directly from the main channel at flood stage.[6] Important features of this connectivity are the surface area inundated, the duration of the inundation, and the depth of the floodwaters. The degree of connectivity is directly related to the microbial productivity of the flood—most especially bacteria—inasmuch as that productivity represents what is available to the entire aquatic food chain network. Rivers might then be compared to one another along a continuum in terms of their degree of connectivity with the floodplain. At one end of the continuum would be a river confined entirely to a main, uniform channel by levees and dikes. Such a river is connected but only within the channel. The current is likely to be swift, and the bacterial productivity—and hence the aquatic life—is likely to be minimal. At the other end of this continuum is a meandering river with a vast floodplain that is periodically flooded and that drains slowly to maximize the nutrients available to the resident bacteria.[7]

The Flood Pulse as Metronome for the Movement of Life

When we think about the movement of a river, the first thing that comes to mind is the vertical flow of the current downstream and eventually to the sea. It is clear, however, that it is the lateral movement of the river at flood stage that represents the most consequential movement for all riverine life-forms. The flood pulse, in this sense, represents the metronome for the entire assemblage of riverine life, shaping the movement of all its species to take advantage of the sustenance it offers. The movement in question is not in the permanent aquatic zone or in the perennial terrestrial zone but in the "in-between" space, the transitional zone that is, by turns, wet and dry. Depending on the hydrology of the particular river, its pulse may vary greatly in amplitude, timing, and duration. The greater its variation, the greater the diversity of habitats that it supports and the greater the adaptability of the species that are found in the floodplain.[8]

The plant life that thrives on portions of the floodplain is quite different than plant life in upland terrestrial zones. In temperate North America, certain trees—willows and beeches, for example—are likely to be found streamside because they prosper on wet, annually flooded soils. Sugar and silver maples as well as green ash are also well adapted to periodic flooding. Most oaks, except for some red oaks, are far less tolerant of flooding. Sedges, reeds, and marsh grasses require a certain degree of flooding because to thrive in the long run they depend on flushes of nutrient-rich water brought by the pulse. Although plants in the short run are stationary, a time-elapse photograph of the flora as the water rises and then ebbs reveals movement that is orchestrated by the pulse metronome. Aquatic seeds and root systems germinate and flourish as the water rises, colonizing the flooded plain, only to die and decompose as the pulse recedes. Conversely, grasses that depend on the rich silt of the floodplain sprout as the water retreats, providing vital fodder for the herbivores, not to mention the pastoralists who move their domesticated livestock in synchrony with the appearance of new grass.

Of all the riverine creatures, the most studied are fish. Their movements are studied by scientists and laypeople alike because they are such a large part of the diet of human populations along the river and often the most important commercial cash "crop" as well. The first thing to note about freshwater fish is that they, like molluscs, crabs, and frogs, are largely confined to the watershed they inhabit. Unlike seagoing fish, such as salmon and eels, most birds, and mammals, they rarely have the option of shifting to another watershed.[9] For this reason, many watersheds have a high proportion of endemic species found in one, and only one, watershed.

Evolving in a unique watershed, many species have adapted over time to take advantage of precisely the characteristics found within it. Along the Amazon, for example, certain fish species have evolved to eat riverine fruit that becomes available only at the height of the flood stage. Other Amazonian predatory fish have adapted to the extreme turbidity of the river by locating prey by means of electric pulses; still others, evolving in oxygen-starved water, have evolved to breathe air at the surface.[10]

Adaptation to the movement of the river, to its unique metronome, is also apparent in the movement of its fish, a large majority of which move within the main channel as well as feed on the floodplain at high water. The purpose of any fish migration is the quest for nutrition, spawning sites, and safety. Some of these migrations span thousands of kilometers and rival those of the better-known intercontinental feats of migratory birds such as the renowned bar-tailed godwit. Fish migrations, for short or long distances, are triggered by environmental signals linked to the hydrology of the river. Such signals can be as varied as water temperature, current speed and volume, turbidity, the scarcity of local nutrition, or the arrival of threatening predators. Some migrations are impelled by the life cycle of the fish species, especially migrations to traditional spawning sites.

The flood pulse, then, though it is not the only cause of fish migrations, is by far the most important for most riverine lifeforms. In terms of pure nutrition, studies have documented that juvenile salmon living adjacent to seasonally inundated habitats have substantially higher growth rates than salmon confined to the channel or to a perennial lake or pond.[11] The superior biotic richness of ephemerally watered habitats as compared with either fully terrestrial or fully aquatic habitats explains the life-giving power of the flood pulse.

There is a vast clockwork choreography of movements triggered by the river. Just as the fish migrate to the tune of the floodplain and its nutrition, so do, for example, waterfowl and other birds move to the floodplain to take advantage of the abundance of fish and hatching insects. The burst of nutrition and growth of mallard ducks provided by the relatively brief winter floods of December and January along the alluvial plats of the Mississippi has been quantified.[12] Dispersing in the shallowly flooded forests of red oaks, the ducks increased their foraging time eightfold and consumed fourteen times their usual intake of animal matter (mostly fish). They rapidly gained weight and fat and, on a daily basis, consumed roughly 4 percent more than required for bare subsistence. The reader can easily imagine the cumulative effects of flood pulse migration: the raptors arriving to consume fish and other birds, the riverine mammals, like otters, foxes, and wolves, coming to prey on smaller birds and stranded fish. The iconic scene of grizzly bears standing midstream to snatch migrating salmon—which form as much as two-thirds of their annual diet—is emblematic of the synchrony of movement animated by fish migration. Virtually all riverine creatures, including *Homo sapiens*, are keyed into the annual succession of movement, especially that of their own sources of nutrition. Thus, fishing folk along the Mekong begin fishing in earnest when certain species of birds arrive because their coming accurately predicts the imminent arrival of the fish that will fill their nets.[13]

Interspecies-Assisted Movement

The synchrony of movement among riverine creatures alerts us that no single species movement can be treated in complete isolation. All that is in motion along the river is best understood as an "assemblage effect" in which the river itself—its path, its temperature, its nutrient load, and its volume—is the key agent. Often the relationship between species is dominated by a predator-prey logic. More often the relationship is far more complicated.

The case of freshwater mussels in the southeastern United States is a striking example of this complexity. Though an individual mussel on its own may live as long as a human, it is "stuck in the mud" (actually, more likely sand and gravel) in its streambed and can move with its one muscular foot only a very small distance. The question then arises of how mussel populations are able to disperse and proliferate throughout the watershed. The answer illustrates the amazing and complex fashion in which interspecies collaboration of a sort arises to facilitate reproduction and movement across the watershed.[14]

Female mussels are fertilized by sperm-casting males upstream from them (as much as ten miles). The fertilized female mussels then breed the eggs as they mature to the larval stage, which may last many months. Larval mussels, termed glochidia, in their thousands infest the gills of the maternal mussel and are visible in the small open space between the two shells. In some species the small aperture closely resembles a small dark minnow, complete with a white eyespot. Seemingly because of this evolved mimicry, a fish will typically approach the mussel to investigate a possible meal. At the moment the fish touches the aperture, the mussel's gills burst in an eruption of glochidia that now, in turn, infest the gills of the inspecting fish. They cling to their host fish for perhaps ten days, at which point those that have survived are viable juvenile mussels. They drop off the host and implant themselves in the streambed. In the course of their journey, they disperse according to the movement and range of their host fish.

Other mussels have evolved a mimicry of an even more astounding variety. The genus *Hamiota* secretes strands of mucus that may be nearly three meters long, at the end of which is a packet of mucus filled with glochidia. As the strand is paid out, it resembles the desired food of its typical host fish. When a fish strikes the "lure," it bursts, and the glochidia attach themselves to the gills of the unsuspecting fish. Once again, the glochidia mature on gills of the host fish until they become viable juveniles and then drop off to colonize a different patch of the watershed. As a virtuoso performance of biomimicry, this has attracted the wonder of scientists. Abbie Gascho Landis writes of the marvel of "blind, brainless invertebrate(s) capturing sperm from a river full of water, acting out an ancient dance to lure just the right kind of fish and then exploding their larvae into the fish's gills to be ferried to new locations."[15]

This charismatic feat of reproduction by freshwater mussels illustrates a more fundamental and consequential truth: the survival and movement of a single species is enabled by much of the biotic assemblage that surrounds it.

Early Humans as River Creatures

Any account of our species' relationship with rivers must begin with the realization that roughly 95 percent of our time on Earth has been spent in small bands of hunter-gatherers. What is striking about hunting and foraging is that it is punctuated by short, intense bursts of activity designed to take advantage of a temporary superabundance of available food sources. Many human hunting activities are calculated to intercept a periodic migration of desirable game: gazelles, pigs, deer, caribou, or buffalo. Such migrations are, in turn, triggered by the abundance of food sources sought by the migrating game: for example, fresh pasture or the mast fruiting of trees and shrubs. In the case of birds, the goal of human hunters may be to intercept migratory species along their flyway, where they rest and may be more easily caught, or at their nesting sites, where they may be killed and their eggs gathered. The birds themselves, of course, are typically on their way to safe nesting and nutrient-rich sites. Perhaps the best-known cases of migration are the periodic mass movement of fish such as salmon, shad, and herring from their freshwater habitat back to their maternal waterway to spawn (anadromous fish) and other species, most notably eels, headed the other way, from their freshwater habitat to spawn at sea (catadromous fish). Humans in many cases are competitors for game and forage with other predators—

for example, with grizzlies for salmon and with other mammals and birds for mushrooms and pine nuts.

For our purposes, these patterns are noteworthy for at least two reasons. First, they are anything but random. It is not as if a forager-hunter sets out daily in the hope that he or she will stumble across valuable game or a lucky patch of mushrooms. Instead, the hunter-forager sets out to intercept an anticipated migration of fish, birds, or game or heads to a corner of the forest where a flush of edible mushrooms has been known to appear seasonally. Bands of hunter-gatherers have, as it were, a collective encyclopedia of the seasonally anticipated appearance of valuable and abundant sources of nutrition. This requires that they be attuned to a large number of distinct natural rhythms that govern the movement and fruiting of each one. Incidentally, it is this strategic interception of migrations and fruiting that helps account for the agrarian/peasant view that hunters and gatherers are shiftless, lazy, and hence primitive. The fact is that, say, two weeks of trapping thousands of eels on their mass migration to the sea or intercepting an annual migration of gazelles can, if successful, provide much of the year's protein. Unlike the peasant, whose work rhythm is more quotidian and unremitting, the intense burst of carefully timed capture and preservation of food yields more in the way of corresponding bursts of leisure.[16]

The second reason this synchrony of coordinated movement is noteworthy is that it largely takes place in that green corridor of life and nutrition that is the river valley and its adjoining floodplain. Early *Homo sapiens* is, in this sense, as much a flood-adapted creature as are the floodplain plants, fish, birds, and mammals. Like other forms of life, humans too tended to cluster in those settings where the density of nutrition was greatest.

A striking instance of this is the Pacific Northwest of North America, where the salmon runs, marine mammals, and large game animals were exceptionally abundant. The region was probably the most densely populated nonagrarian (hunter-gatherer) setting in the world. Its natural affluence is attested by the unparalleled florescence of material culture, not to mention conspicuous consumption in the form of the potlatch as a political strategy.[17] At the base of the pyramid atop which perched the Pacific Northwest tribes was a rarely equaled concentration of organic nutrients that sustained the vertebrates: the fish, birds, and mammals that represented a vital portion of the top predators' diet.

Sedentism and Rivers

With permanent settlement, usually linked to fixed-field agriculture, come the first tentative steps in changing the river rather than adapting to it. Eventually, and usually with unintended and often catastrophic consequences, early humans did have an impact on the liquid landscape of the river. I refer to this long epoch as the "thin Anthropocene." Thin in two ways: first, because there were so few of us on the planet, and second, because we had so few tools with which to reconfigure the landscape. This division distinguishes a long period of modest human impact from the thick Anthropocene, ushered in by the industrial revolution, that put the machinery and technology in our hands to remake the landscape.

Virtually all the earliest recorded city-states were located on floodplain (alluvial) lands. It was possible, and in fact reasonably common, for there to be fixed settlements of some size coupled with grain cultivation that did not coalesce into city-states. But a city-state without a concentrated grain-growing population base was extremely rare.[18] Only rich, annually renewed alluvial soils could, given the constraints of transport in the ancient world, provide the concentration of manpower and taxable, storable foodstuffs that made even modest state-making possible. Just how modest deserves emphasis. The typical early city-state in the Mesopotamian alluvium had fewer than twenty thousand subjects and could be crossed on foot in a single day or less.

The impression that such states were based on irrigation works constructed by a state-mobilized workforce, most famously championed by Karl Wittfogel in *Oriental Despotism,* has been proven mistaken in almost every respect. Quite the contrary. The most common form of early agriculture is known as "flood recession agriculture" (in French: *cultivation décrue*). It is still practiced throughout the world because it has been shown to be the most labor-saving form of agriculture, regardless of the crop being planted.[19] The flood does almost all the work. It first drowns virtually all the competing vegetation and then, if well behaved, lays down a nutritious layer of fine silt, leaving behind a perfectly harrowed field for the cultivator to sow. The fertile silt in question is, of course, a transfer by erosion of upstream soils. All that remains for the cultivator to do is broadcast or insert seeds in the prepared soil, giving the crop a head start on other plants. The humans who practiced this form of agriculture in the Nile Valley and along the Euphrates and Yellow Rivers were a flood-adapted species, quite like the birds and mammals who came to feast on the offerings of the flood pulse.

The extensive floodplain of the Ayeyarwady River in Burma exemplifies the relationship of early settlement and the river.

The earliest agriculturists in the Ayeyarwady Basin were the Pyu people in the middle of the first millennium CE, who settled along the smaller side valleys of the Ayeyarwady where they could benefit from a gentler flood pulse but not risk a catastrophic inundation.[20] They practiced a form of irrigation that could be easily implemented by constructing small weirs to divert high water to cropland and ponds. We might think of it as a locally modified flood pulse. That their endeavors were local and autonomous is made clear by the fact that they were in place well before small states were formed and continued well after those had collapsed. Much the same is the case for the Burmans who succeeded the Pyu. The heartland of their agricultural production was also located in the Dry Zone. The Dry Zone lies in the rain shadow of the Arakan Mountains (Arakan Yoma), but the monsoon rains outside the zone feed a good many perennial rivers and the floodplains adjoining them within the zone. All that is generally required is to carve a small opening in the natural levee of such a perennial river at high water to sufficiently flood the padi fields for planting. In this case, it is the water that brings most of the nutrition to the crop. The small weirs and canals that conveyed the water to the major padi crop were sown in small nursery plots just before the monsoon rains in June and then transplanted to the flooded fields in late July or early August. The intervention into the riverine landscape was minimal and easily managed by cooperative village labor.[21] Human landscaping, albeit on a minute scale, had begun. Its transformative effect is still modest, even compared to that other mammalian landscaper of rivers: the beaver. At this stage, humans were largely adapting to the movements of the river; they were under no illusion of being its master.

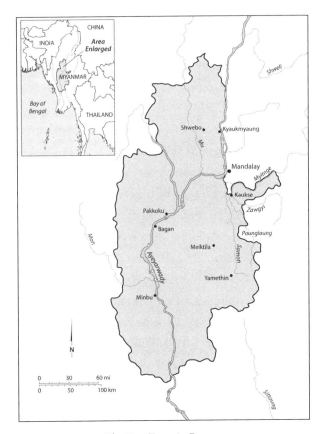

The Dry Zone in Burma.

Early sedentary communities were "river creatures" not only because they depended on silt-laden water to nourish their fields. They were also reliant on navigable water, at least occasionally, because, as denizens of the lowland floodplain, they were far from self-sufficient economically. Rich as their immediate habitat may have been in soils and wetland abundance, they required products from other, typically upland zones that they could only acquire by trading. Most often the geographical border between lowland agrarian communities and their complementary trading partners was the fall line, sometimes called the piedmont (a corruption of the French *pied du mont*—that is, foothills). The fall line marks the beginning of a transition to an upland ecology and, significantly, the point at which upstream navigation becomes difficult or impossible. Sedentary communities located at the fall line often become important trading and transshipment points and, much later but before steam power, centers of early water-powered flour and textile mills. Throughout the ancient world, these locations were of strategic military and economic importance. Control of these choke points conferred command of the upstream and downstream traffic. The classic riverine Malay settlements were configured by an upstream/downstream (*hulu/hilir*) pattern in which the center of power derived from ruling over the fall line. In a similar fashion, the Celtic trading oppida at the fringes of the Roman Empire were located at the fall line or choke point of one or more rivers.[22]

The centrality of navigable water for early sedentary communities in the floodplain derives almost entirely from what is called the friction of distance: the fact that movement over water is vastly more efficient than over land. A striking case in point is that as late as 1800 (before steamships and railroads) it was about as fast to go from Southampton, England, to the Cape of Good Hope by ship as it was to go by stagecoach from London to Edinburgh! And, of course, the ship could carry far more cargo.[23] Transport by water in preindustrial Europe was estimated to cost one-twentieth of overland transportation. Despite the fame of the roads in the Roman Empire, Diocletian's Price Edict recorded that the price of a wagonful of wheat doubled every eighty kilometers it traveled. Overland shipments of coal in the sixteenth century were calculated to lose 10 percent of their value per mile, making long-distance trade prohibitive. Grain shipments, being more valuable per unit of weight and volume, could be made up to 250 miles before they became a losing proposition. Exceptionally valuable goods might well justify long overland trade (the Silk Road), but bulk commodities, many of them essential to lowland communities, were only economical if they traveled by water.

Water and waterways, then, provided they are navigable, connect human communities in ways that extend well beyond the exchange of goods. They represent a zone of contact, intercourse, and cultural influence that, over time, fosters a cultural homogeneity despite political rivalries. The most influential exploration of this process is to be found in Fernand Braudel's *The Mediterranean World*. Braudel shows how the navigable Mediterranean, thanks to long and intense commercial and cultural connections, came to represent a recognizable world of mutual influence—in cuisine, urban structure, and political forms. Much the same case has been made for the Sunda Shelf of Southeast Asia, where trade and migration were, if anything, even easier than in the Mediterranean. This integration distinguished the maritime peoples from societies in the hilly regions not far from the coast that were, by contrast, far more distinct from the coastal and riverine settlements. In terms of contact and familiarity, peoples who are, say, separated by three hundred miles of easy water are, in most meaningful senses, closer and more integrated than people who are only thirty miles apart but separated by rugged mountain terrain.

The integrative effect of a major navigable river is exemplified by the Ayeyarwady in Burma. Since the arrival of Burman military settlers from the China-Tibet border area in the first millennium CE, the river and its floodplain have come to represent the heartland and "superhighway" of Burman culture. The river's cultural effects are in no small degree the result of its navigability for seventeen hundred kilometers from the upstream post of Bhamo to the head of the delta just north of Hinthada. Along this great stretch of the river, one encounters a largely padi-growing Burman population, speaking the same language, professing a recognizably familiar form of Theravada Buddhism, and observing a common set of seasonal rituals, myths, dance traditions, marriage rites, and burial practices. Travel was, until the twentieth century, centered almost exclusively on the Ayeyarwady. In the British colonial era, the river supported what has been described as the largest freshwater fleet in the world, the Irrawaddy Flotilla Company, which, at its peak in 1920, operated six hundred vessels ferrying cargo and passengers upstream and downstream. But well before the colonial era and steam power, the navigable water of the Ayeyarwady made possible a form of transport and commerce that would otherwise have been inconceivable.

The Irrawaddy Flotilla Company was simply an industrial amplification of a symbiosis with the river that had been in place for as long as there had been *Homo sapiens* along its banks. While the particulars of this human symbiosis may differ from that of other mammals, fish, molluscs, and birds, it bears an undeniable family resemblance. For humans, as for fish and birds, the river is a corridor and a habitat. Like birds, fish, and riverine mammals, humans in this niche interbreed, both in the literal genetic sense and in the sense of cultural, social, and linguistic *speciation*. The cohesion of Burman ethnicity is largely a product of the river. What the river provides, beyond a natural concentration of

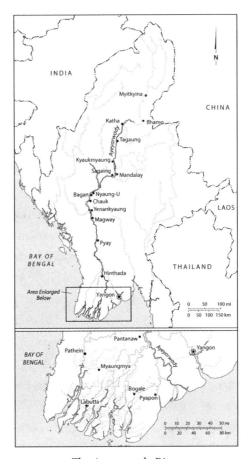

The Ayeyarwady River.

available nutrition, is relatively frictionless movement that presents few obstacles to passage. Just as the river water is the niche and highway for migrating fish and the flyway above the river is a corridor for birds, so the surface of the river and the adjacent floodplain provide a socially integrative corridor for humans. The movement of the river—seasonal and sporadic—dictates the ecology to which all these species must adapt.

When the cargo itself floats, no freight vessel is required. The historically most important commodity of this kind, still important today, is wood, specifically bamboo and logs. Much of the bamboo is cut along a tributary, the Shweli, and lashed together by rattan into huge rafts that make their way slowly downstream to ports like Mandalay, where the rafts are dismantled and the bamboo sold. Teak logs, the most valuable wood, are less buoyant, and they are either slung under the bamboo rafts, both to buoy and conceal them from customs officials, or else carried in large barges built explicitly for that purpose.[24]

More impressive still is the way in which huge ("Alibaba") jars, many of them turned and fired in the town of Kyaukmyaung near clay deposits, are floated in their thousands downstream after being lashed together into makeshift rafts. The jars themselves provide the flotation; their mouths are situated upright and slightly to the stern (upstream) so that they do not take on any water under choppy conditions. How else such large and relatively fragile jars could be transported to markets far away under premodern conditions is hard to imagine. It would require hundreds of slow-moving ox carts with a great deal of straw insulation, and even then many would certainly crack, not to mention the expense of draft power and cart drivers. The final advantage is that, at the end of the journey, there is no vessel to dispose of or to return back upriver to Kyaukmyaung.[25]

Given the fact that the fall line typically represents the border between quite distinct ecological zones, the populations of each zone have a great deal to gain from access to products from the other. That access can come from raiding and/or trading. Of the two alternatives, trading is the steadier and more reliable. Trade, in turn, can come by water (river, sea, ocean) or overland. The

Men lashing jars together to make a raft.

decisive difference, as is apparent from the examples of bamboo and ceramics, is that exchange by waterborne trade is so much more efficient that it opens the gates to a vastly larger spectrum of commodities.

The logic of transportation is, as noted, quite straightforward. Goods with highest value per unit of weight and volume can be shipped very long distances and still justify the labor and time required. Thus, goods such as refined gold and silver, obsidian, gems, spices, rare medicinal herbs, and rare aromatic woods were moved across great distances overland in the premodern world. Depending, again, on their value at the destination, upland products in mainland Southeast Asia such as gaharu wood, rhinoceros horns, candlewood, edible bird's nests, beeswax, honey, rare feathers, rare resins (for example, tung oil), tea, tobacco, and opium might warrant overland transport. The calculations governing overland trade involved the ruggedness of the terrain, the danger of robbery, and the availability of pack animals or human

The resulting raft composed of jars.

porters. Carts allowed more freight but required infrastructure in the way of available fodder and viable cart tracks.[26]

For the trade in livestock, the logic is different, inasmuch as some livestock can travel (or be driven) overland under their own steam. This logic can be extended to captives and slaves. Slavery was, in fact, ubiquitous in Southeast Asia. Human "livestock," often captured at higher elevations in the watershed, was among the most important trade commodities, the journey typically descending along watercourses from the hills to the river valleys, mimicking, as it were, the movement of the water itself. Slave trading represented another instance of how the river shaped the movement of people and commodities—and of people *as* commodities.

THREE

Agriculture and Rivers
A Long History

> Man is everywhere a disturbing agent. Wherever he plants his foot, the harmonies of nature are turned to discords.
> —GEORGE PERKINS MARSH

Rivers as corridors of travel and nutrition were no less vital to our hunting and foraging ancestors, who represent all but the last ten thousand years of our adventure as a species, than they would become for the early civilizations that were invariably founded on riverine floodplains. And yet, the hunter-gatherer imprint on the hydroecology of the rivers along which they dwelled was minimal. They took full advantage of the natural abundance of the riverine landscape, but they largely adapted *to* the movement and pulse of the river rather than altering the landscape. Their numbers were comparatively small, and the only powerful tool at their disposal was fire. They used fire to resculpt the riverine flora to encourage desirable plants, berries, and nuts and, above all, to create the browsing landscape that would attract prey. Much of the animal protein consumed by hunter-gatherers was either intercepted by exploiting the predictable migration routes of prey or harvested from a fire-induced landscape designed to draw prey to it. The use of fire in this fashion created patches of open land where a new floral succession could begin. It increased the biodiversity of riverine ecology but made little difference in the "natural" movement of the river.

Sedentism and fixed-field agriculture, together with the population growth they sparked, marked the epochal change in humankind's relationship to the river, and the beginning of what I call the thin Anthropocene. The thin Anthropocene stretches from roughly 8000 BCE, giving way to the thick Anthropocene beginning around 1700 CE; the latter era is marked by an exponential growth of the human population and, beginning about 1800, the invention of powerful new tools to modify the environment—tools we associate with the industrial revolution.

If we wish to see, in embryo, what humankind has done to the riverine landscape and watersheds in general, it was all on display throughout the thin Anthropocene, although the pace of transformation was slower and the imprint fainter than the massive landscape modification made possible by the industrial revolution. Cumulatively, however, the impact was massive. In the comparatively short time of ten millennia, agrarian *Homo sapiens* had revolutionized, often inadvertently, the hydroecology of countless rivers to a degree that dwarfed what their hunter-gatherer forebears had achieved over roughly two hundred millennia.[1] A much closer look at their interventions and intentions reveals a desire to simplify and tame rivers that would be more fully and radically expressed in the thick Anthropocene.

Sedentary, fixed-field agriculture typically requires the removal of vegetation that competes with the desired crop for sunlight, water, and soil nutrients. Hence, it mandates both deforestation to create an arable field and, as the crop ripens, the removal of smaller but more numerous competitors: weeds. It follows that, as agriculture expands, it leads to the contraction of forests. Since so much of early sedentary agriculture is located on the floodplain, the floodplain-adapted forests were the first to go.

If deforestation was merely the result of clearing land for cultivation, it would be confined to the areas sown to crops. The appetite for wood in agrarian societies, however, is far vaster and insatiable. First, and most obviously, wood is needed for the construction of permanent housing (the hearth, or *domus*) and for granaries and fencing to confine and protect domestic animals and crops. A signature feature of virtually all sedentism is its function as a refuge for domesticated plants and animals that can only survive if protected and defended by humans. A substantial part of that defense requires wood. Fuel for cooking and warmth also requires firewood. Hunters and gatherers, of course, also cook and warm themselves with fire, but what is distinctive about sedentary communities is that the firewood must be gathered from in and around the settlement itself. As the nearby firewood is depleted and the cost in time and labor of hauling it over longer and longer distances rises, there are, under premodern conditions, only two prominent solutions, short of relocating the village. One is to make charcoal by a controlled burn of wood where it is felled. Although it is far more wasteful of wood, charcoal provides more heat value per unit of weight and volume than raw wood and can therefore be transported over longer distances economically. The second solution is to drastically reduce the transportation cost of moving raw wood by floating logs down to the settlement from upstream in the watershed. Both solutions have consequences we shall examine later in this chapter.

Growing floodplain settlements ushered in a cascade of new technologies that were unprecedented in their appetite for fuelwood. Pottery, for example, is strongly associated with sedentism and cereal grains. Various pots are used for storing grain and water, cooking, and fermentation, and the most durable pots are fired in kilns that require high temperatures and hence a great deal of firewood and/or charcoal. Much the same is true for the bricks that are used to build durable elite and ritual structures, not to mention the walls that protect larger settlements from raiding and sieges. Some walls are built with sun-dried bricks, but a great many are kiln fired.[2] The Bronze Age and the Iron Age are virtually defined by the smelting of ores and hence by massive requirements for fuelwood and charcoal. A part of early metallurgy was integral to the expansion of agriculture in terms of plow tips, durable cart parts, threshing knives, and scythes. As sedentary communities grew and morphed into city-states with hierarchies, monarchs, and armies, the demand for fuelwood expanded greatly, much of it devoted to the production of early military hardware: armor, swords, spears, and arrow tips.

All these firewood-intensive technologies were well established before the Common Era, and collectively, as sedentary populations grew, they represented an assault on ancient forests. As premodern centers of population exhausted the timber supply closest to the core settlement, they began logging farther upstream, cutting timber along the banks, and taking advantage of the buoyancy of wood to float logs down to their settlement. Minimizing the labor required dictated that trees be felled as close to the river as possible. As the nearby upstream banks were deforested, additional fuel had to come from farther and farther upstream and/or from smaller trees that could be more easily transported to the river's edge. Where the flood pulse was pronounced and the floodplain gradient low, trees could be felled in the low-water season and then launched downriver when the rising waters reached them.

There is abundant evidence of deforestation in the classical world, from the Athenian quest for naval timber in Macedonia to the shortage of timber in the Roman Republic.[3] Much earlier, by 6300 BCE, in the Neolithic town of Ain Ghazal, there were no more trees within walking distance of the settlement and firewood had become scarce. As a result, the community dispersed into scattered hamlets, as did a good many other Jordan Valley Neolithic communities when the population exceeded the carrying capacity of their local woodlot.[4] *The Epic of Gilgamesh* beautifully encapsulates the close connection between the founding of the earliest agrarian states and the deforestation of the watershed. Its hero, Gilgamesh, slays the giant who guards the great forest and builds a raft with its timbers to return to the land where he will found a city; the timbers from the raft are then fashioned into the central gate of the new kingdom.

Nowhere has the deforestation of a major watershed been so catastrophic and so thoroughly documented as in the case of China's Yellow River. The river itself, of course, acquired its name from the color of its sediment-laden waters. Along with the Amazon, the Brahmaputra-Ganges, and the Ayeyarwady, the Yellow River ranks among the most sediment-clogged in the world.

The increase in the sediment carried by the Yellow River is historically closely tied to the destruction of the watershed flora. That deforestation is, in turn, a direct result of the clearing of

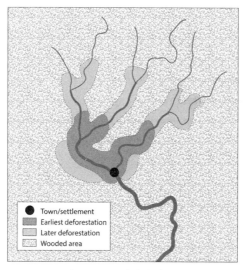

Pattern of upstream deforestation from a hypothetical premodern population center.

land for agricultural settlement corresponding to military expansion and the forced dispersion of nomadic herding populations. The sediment carried by the river came largely from the Ordos Loop and the loess soils through which it passed. Loess soils are homogenous, silty, rich in minerals, and easily worked. In short, it is the ideal soil for premodern cultivation. At the same time, it is easily erodible, particularly when exposed to sun and rain on steep hillsides.

Paleogeographic research allows us to reconstruct the fluctuations of the river's sediment load over time. In keeping with our emphasis here on a significant thin Anthropocene long before the last two and a half centuries of river engineering, the damage to the Yellow River's watershed began more than seven millennia ago. It increased markedly between three thousand and two thousand years ago as Han agricultural colonization, with military support, promoted more intensive cultivation. A final premodern burst of erosion and flooding dates from roughly 1050 CE through 1280 CE (Song Dynasty), marked again by agricultural expansion, military construction, charcoal production, and brick and papermaking.

The cumulative effects of stripping the watershed of its earlier forests were tragic; the Yellow River became known as China's Sorrow. From the eleventh to the thirteenth centuries CE, the channel of the river radically changed course eight times, moving north and south of the Shandong Peninsula. This was a direct result of the accumulation of sediment deposits on the flat North China Plain that clogged one channel after another. The sudden leaps of the river to carve a new channel meant death and destitution for those peasants unwittingly cultivating the floodplain in the path of the increasingly erratic river. Nor was the river the same hydrological entity it had been before the loess plateau was stripped of most of its earlier vegetation. No longer were rains and snowmelt absorbed by forests, now felled, and wetlands, now drained. The result was an unprecedented and rapid surge of floodwaters further eroding the upstream watershed and carrying ever greater amounts of sediment downstream. Deforestation had the additional effects of raising the ground temperature, hastening evaporation, and lowering the water table. For those cultivating the loess plateau it meant the loss of many springs and a greatly increased chance of periodic crop failures and famines.

The reader may have noticed that the hegemonic narrative around ancient deforestation is resolutely centered on its effects on humans and sedentary agriculturalists in particular. These effects were, to be sure, devastating. They were unexpected as well. The earliest centers of sedentary agriculture, whether in Mesopotamia or along the Yellow River, could hardly have anticipated the far-reaching consequences of their interventions in the watershed. And when the effects did become apparent, they were difficult to reverse for both ecological and political reasons. Far from being deplored, however, deforestation was seen at the time as a triumphal signpost in the creation of agrarian civilizations. This passage from the writings of the classical scholar Mencius captures—indeed, celebrates—the subduing of a disorderly nature as necessary to the creation of a superior, human-designed order.

> In the time of Yao, when the world had not yet been perfectly reduced to order, the vast waters, flowing out of their channels, made a universal inundation. Vegetation was luxuriant, and birds and beasts swarmed. The various kinds of grain could not be grown. The birds and beasts pressed upon men. The paths marked by the feet of beasts and the prints of birds crossed one another throughout the Middle Kingdom. To Yao alone this caused anxious sorrow. He raised Shun to office, and measures to regulate the disorder were set forth. Shun committed to Yi the direction of the fire to be employed, and Yi set fire to, and consumed, the forests and vegetation on the mountains and in the marshes, so that the birds and beasts fled away to hide themselves.[5]

From a naturalist's perspective, this and most hegemonic human-centered narratives ignore the fate of the myriad of nonhuman life forms—a glaring omission. The Mencius text at least has the virtue of noting that human settlement necessitated the destruction or banishing of a natural world that stood in the way, even as it praises the victory of human-imposed order on the "disorder" of the natural world.

Insofar as the early human impact on the watershed was narrated at all, it was of course narrated by humans. To the degree that this account took note of the impact of human intervention on other species at all, it was largely confined to those effects that most directly affected the material welfare of humans: a lower water table, the decline in valued fish populations, the increase in damaging floods, and the relative absence of once abundant populations of game birds and mammals.

The pace of humankind's transformation of the riverscape before, say, 1750 CE was essentially determined by the growth of the human population (see the table) and the spread of fixed-field agriculture (domesticated plants) and pastoralism (domesticated animals). In keeping with the trajectory of population growth, that pace accelerated slowly at first, gained speed in the first millennium CE, and then rose exponentially in the industrial era.

ROUGH ESTIMATES OF WORLD POPULATION

1 CE	250 million
1000 CE	300 million
1750 CE	750 million
1900 CE	1.5 billion
1950 CE	2.5 billion
Present	8.0 billion

The choice of 1750 as a significant date is predicated on the fact that it predates the advent of steam power, the explosion of engineering technology that, for the first time, enabled massive interventions in the riverscape.

We have no reliable catalogue of what this early era meant for nonhuman species. We can, however, venture some educated guesses about the impact based on what we know about the gross changes in landscape brought about by agriculture, pastoralism, kiln drying, and metallurgy. The cumulative effect of this series of interventions was to simplify and homogenize the watershed and reduce the biodiversity of its habitats and resident species. The resulting new assemblage of species was confined largely to those best adapted to the new anthropogenic conditions.

Consider, for example, the floodplain and banks of a watershed now populated by cultivators and pastoralists. The trees and much of the vegetation has been cleared and replaced by a smaller number of sown crops and/or grasses that can withstand periods of browsing by domesticated herds and flocks. A diversified range of riverine habitats suited to a comparatively wide range of birds, mammals, insects, and smaller organisms has been reduced to a narrower, far simpler habitat. This more uniform habitat could not accommodate anything like the biodiversity of its predecessor. It would be a mistake, however, to see the earlier landscape as alive and the new one as dead. Rather, it is as if a great many of the previous inhabitants had been evicted or fled from their domicile, leaving behind residents that adapted to or were attracted to what the new habitat had to offer. One assemblage of species has been replaced by a new assemblage, with some overlap. What is certain, however, is that in keeping with the more uniform habitat, the new assemblage would be less diverse than the previous one. It might contain even larger numbers of those species it *does* accommodate—birds and rodents attracted by grains and legumes, soil organisms adapted to disturbed, plowed fields—but the new assemblage would be far more homogenous.[6]

The drainage and diking of agricultural lands are as momentous as deforestation in their effects on the riverine environment. A flood pulse regime, left to its own devices, is a powerful generator of diverse habitats. As it rises and recedes, it recharges the wetlands with moisture and sediment, leaving behind marshes, swamps, fens, bogs—landscapes that are perennially wet, those that, like vernal ponds, are ephemerally inundated, and those habitats that are flooded only during the flood pulse itself. In many undisturbed riverine contexts, much of the watershed is dominated by such wetland floodplain landscapes. They provide, especially in flat, low-gradient terrain, an enormously varied range of habitats together with the species that inhabit them.

Along with clearing the land for crops (deforestation), the second founding act of floodplain agriculture is drainage. This is, in effect, the war *to exterminate mud* and replace it with well-drained soil. What remains are watercourses (drainage and irrigation ditches) and arable land. It represents a boon for crop-planting *Homo sapiens* but a cataclysm for most of the species that dwelled in the variegated wetland habitat. Ellen Wohl and other environmental scientists have called this long-term process the "Great Drying."[7] It is a war not only on mud but on shade: a shade that protects mud and moisture generally. The deforestation of the floodplain and riverbank, so that the open, cultivated fields welcome full, ripening sunlight, is the initial engine of the drying process. The next and decisive step is the shunting of excess water back to the river with drainage ditches and dikes that raise the soil temperature, thereby adding further evaporation and hastening the pace of the Great Drying.

Ironically enough, humans, the agents of this environmental transformation, were largely unaware that they were undermining the physical and organic basis necessary for agriculture itself. The very soil on which they relied for their new subsistence was the product of centuries-long processes of wetland plant growth and decomposition. By interrupting that process, they not only created a more impoverished habitat, inhospitable to many species, they were depleting the natural capital on which their immediate prosperity depended. This depletion accelerated mightily with the industrial revolution, but the template for its beginning was firmly in place more than four thousand years earlier.

The hydrological transformation initiated by agriculture and its second- and third-order ramifications are so numerous as to defy careful accounting. Behind all of them, however, is a straightforward commonsense logic. The more uniform and homogenous the riverscape, the less diverse the species that inhabit it. Neolithic landscapes are engineered for a single purpose: to satisfy the taste and needs of a single species. If the desired landscape requires deforestation, drainage, and domesticated animals, it limits or excludes those species that require trees and wetlands to thrive—forest birds, waterfowl, amphibians, turtles, water snakes, and riparian browsers. Should, as is often the case, the cultivation and deforestation fill the river with unprecedented amounts of silt, this will systematically favor silt-adapted species (for example, certain molluscs, catfish, and carp) and disfavor aquatic insects whose larvae it smothers and fish that require a clear pebble or gravel backwater to successfully spawn. If drainage and rapid runoff from denuded banks make for a much faster current, this will systematically select against those species of fish that flourish best in backwaters and eddies. A river modified by centuries of agricultural settlement is itself likely to be radically simplified; it is likely to consist more of larger channels and main tributaries and less of swamps, wetlands, and marshes. As a simpler, more abridged river, it is necessarily host to a simpler, more uniform and abridged roster of nonhuman inhabitants.[8]

Interlude
An Introduction to the Ayeyarwady

> [Many men] would find it hard to make out as good a claim to personality as a respectable oak tree can establish. [The reader is invited to replace "oak tree" with "river."]
> —GEORGE PERKINS MARSH

Local Spirits, Local Voices

One central purpose of this book is not only to recognize the animated liveliness of the river and its tributaries, but also to give voice to all the flora and fauna whose lifeworld centers around the river (see chapter 5). Their voices are not simple to discern and express, but their needs, the conditions necessary for their flourishing, are largely known. They, of course, represent the overwhelming majority of riverine citizens, but their fate is beyond their control.

For most human citizens of the Ayeyarwady, the landscape of the river, both aquatic and terrestrial, is also "peopled" by a great variety of spirits (*nats*). These major actors, embedded in local history, possess superhuman powers, both benign and malicious. They are, by turns, worshiped, placated, evaded, and invoked. There is little doubt that they exert enormous power over the fate of all riverine creatures. They are, of course, absent in "scientific" studies of the hydrology, geography, and geomorphology of the river and in secular studies of the river's citizens.

And yet, no holistic study of the river can ignore them. They are as integral to the landscape as the local gods that animists discern in trees, mountains, and streams, or as the local gods that persist (however subordinated) in the fabric of peasant Catholicism, village Hinduism, and rural Islam.

Two of my Burmese collaborators have conducted something like a census of *nats* in the upper Ayeyarwady Basin. I present their work here. Not only were they far better equipped than I to conduct such a survey, but I feared that if I had accompanied them, the villagers with whom they spoke would be less forthcoming, anticipating skepticism, if not contempt, for their "superstitions."

The narratives gathered by Maung Maung Oo and Naing Tun Lin follow.[1]

The River and Spirit Worship

During natural and man-made disasters, Theravada Buddhism offered spiritual solace but did little to alleviate immediate crises. In these moments, the *nats* were beseeched to aid local communities. These deities continue to be honored with feasts and festivals, demonstrating their enduring relevance.

Unlike gods in many other cultures, the *nats* in Burma have a unique origin. They were once human beings who, due to their life experiences and the locales where they lived, became revered spirits. A few individuals ascended to the status of *nat* associated with the river, a spirit deemed worthy of homage by believers.

CONSIDER THE STORY of the Maha Giri siblings, whose narrative originates in the kingdom of Tagaung, believed to be the first settlement of the ethnic Burmese people, according to Burma's oral history. In the year 900 CE, during the reign of King The-Le-Kyaung in the kingdom of Bagan, the area that is present-day Tagaung was ruled by Tagaung Mingyee. In Tagaung, a blacksmith named Maung Tint-Te was so formidable that his hammering could create tremors felt throughout the entire city. Fearing the blacksmith's strength and potential challenge, the king lured him by promising a royal connection and a noble title. This was a ruse, and Maung Tint-Te was captured. As he was resistant to weaponry, the king ordered him to be burned at the stake near a saga tree.

Maung Tint-Te's sister, unable to bear his fate, leaped into the flames to join him. The brother and sister became *nats* and were soon known for causing malicious mischief around the saga tree. In response, the king had the tree uprooted and cast into the Ayeyarwady River. As the tree floated downstream and reached Bagan, the spirits implored the king of Bagan to create a dwelling for them, as they remained attached to the floating tree. The king had the tree sculpted into the figures of the spirits and constructed a shrine for them. In gratitude, the spirits vowed to protect the king and his subjects. Citizens were encouraged to build small shrines for the *nats* in front of their houses; eventually the tradition became to erect these shrines in a corner of the living room. The *nat* festivals are celebrated in the Burmese month of Nadaw (December), and the brother's name has been altered to Min Maha Giri, signifying the great mountain king guarding households. Even in the palace of the last Burmese king, a shrine was dedicated to Min Maha Giri *nat*.

The list of *nats* continues to expand, with new spirits being added, often individuals of historical significance or those whose untimely deaths left a mark in the hearts of the public. At times, even ruling monarchs align with their subjects in appealing to the *nats*. There are also local *nats* with vague or unknown origins, yet they are revered in their respective localities, where people pay homage to them in hopes they will satisfy various needs.

Notably, there are three kinds of *nats*. First, the thirty-seven officially recognized *nats* listed in the dynastic chronicles that are traditionally respected by the populace. Second, the Taung Byone brothers, two *nats* known nationwide, whose worshipers gather every August in Taung Byone Village north of Mandalay. Since colonial times, even governments have become involved in managing this event by providing transportation and security as well as making sure that the wildly nonconformist festival (cursing authorities, drinking, and sex, comparable in some ways to black masses and Carnival in Catholic societies) doesn't threaten political authority. Finally, there are historical figures in different regions who met unusual fates and, through oral traditions and stories, have come to be regarded as *nats*. Though they may not appear in the chronicles, they enjoy widespread respect and veneration.

The third category comprises local *nats* known as Ywa-Daw-Shin or Ne-Daw-Shin, meaning the Lord of the Village or the Lord of the Circle. Their origins are often shrouded in mystery, and they are typically depicted as elderly men dressed in white, referred to as Bo-Bo-Gyi. The transition from human to *nat* is not always clear, and sometimes these spirits are named Ah-Me or Me-Daw, signifying "mother," or Ah-Ma-Daw for an unmarried spirit. Married *nats* bear the title Ga-Daw, meaning "Madam" or "Dame." Some *nats*, known as La-Mine, remain enigmatic in name and type but continue to be important to local believers. Rituals appealing for their help typically involve offerings such as three combs of bananas and a coconut placed in a tray or bowl. These smaller *nats* are believed to aid people in their daily endeavors, shielding them from obstacles when harvesting crops, hunting in the jungle, or fishing in rivers, lakes, and fish ponds. The belief in these *nats* persists even when the individuals who venerate them move to a new region.

Local *Nats* at the Confluence

Originating in the far north, two rivers, the Mali Kha and the N'mai Kha, flow southward, converging to create the Ayeyarwady River. This confluence, known as Myitsone, lies at the heart of the native peoples' heritage, where history, culture, and traditions have taken root.

DAW MARIP LU RA, a farmer from the village of Tanghpre, situated eight miles downstream from the confluence, shared a story related to Myitsone. According to local folklore, the most formidable deity in the area was In-Khaing-Bum, notorious for his volatile temper. His anger often led to disasters like landslides and floods. Once he unleashed his wrath on a wooded area with thunderbolts, leaving trees crushed and eroded along the riverbanks. The locals dared not linger in the vicinity, fearing that they might offend his spirit. A shrine was constructed near the confluence to pay homage to this deity.

Traditional practices persist to show respect for such river deities.

THERE IS A TALE of two spirits, Karin Naw and Karin Gam, who vanquished a troublesome serpent spirit that plagued the locals' livelihood. The two brothers used the Kim Gi Nam plant to defeat the dragon. Even today, locals carry this plant on their river journeys to safeguard against potential dangers. When severe floods or other river-related calamities occur, locals believe that their actions might have provoked the deities, thus resulting in havoc. Rituals may vary, but people of different faiths adhere to these traditions to appease the spirits.

Pyin-Daung-Lay Mae Daw is a revered *nat* in the region of Pyin Lay Pin Village, near Inn Ywa Village in the Katha District. The village is situated near a flooded lake known as Pyin-Daung-ay, where the locals rely on fishing for their livelihood. Although the history of this *nat* remains untraceable, it's intriguing to note that some of her worshippers, whose fish-farming business declined, migrated south to the delta region and transported reverence for this *nat* to various villages along the waterways.

IN THE VICINITY of Doe-Bin Village, nestled between Inn-Ywa and Nga-Oh Villages, there stands a *nat* shrine dedicated to the sibling *nats* of Kyun-Pin. Legend has it that these brother and sister *nats* were once descendants of the Sawbwa of Mong Mit.

Their transformation into *nats* occurred during a riverside bath; they tragically lost their lives when a massive teak tree collapsed on them. The locals believed their spirits clung to the tree upon their death. This tree, carried by the currents, journeyed from the Shweli River to the Ayeyarwady, covering a distance of about 180 nautical miles, until it reached Mingun, across the river from Mandalay.

In Mingun, these *nats* appeared in the dreams of a local Buddhist monk, Oak-Chut Sayadaw, who was deeply engrossed in his meditative solitude. They expressed their desire to reside near the Mong Mit Hills, as these hills were historically connected to the Prince of Mong Mit, a son of King Bodawpaya, and one of his queens born to the Sawbwa of Mong Mit. Responding to their call, the mendicant had a small shrine constructed between two grand teak trees at the foothills.

As rafters and travelers navigated the river, they started paying homage to these sibling *nats* when they approached the shrine. Initially, they would invoke the *nats'* names and request assistance in navigating potential river hazards such as rocks, sandbanks, and submerged debris. Over time, stories of these *nats* safeguarding and aiding the worshippers who sought their favor made them highly revered throughout the region. Consequently, the humble birdhouse shrine was transformed into an established altar dedicated to these *nat* siblings. The new sanctuary built by the worshippers later surpassed the original shrine located near the banks of the Shweli River in Katha Township.

The rituals involved in honoring these spirits closely resemble other such rites. Offerings like the Ga-Daw-Pwe, comprising three combs of bananas, a coconut, incense sticks, candles, and customary items, are presented. Locals refrain from engaging in inappropriate behaviors in the vicinity of the shrine such as urinating, as they believe such actions could provoke the ire of these deities. Furthermore, it's a common practice for couples to avoid public displays of affection, as this could potentially disturb the *nat* siblings residing there.

Downstream Singu

There are patterns of *nat* worship along the river downstream from the town of Singu in the Mandalay Region. U Aung Thin, a fisherman in his sixties native to the area, was asked to describe the *nat* worship practices in the region. What follows is woven from his narrative.

> AS THE AYEYARWADY flows past the third defile, the river spreads out to form a floodplain where the locals work the exposed riverbed to grow seasonal crops. What they grow differs according to the distance and height from the river. Most of the locals are Buddhists, so rites and rituals are mainly centered around Buddhist and related animist worship. Almost every family that has a house has a shrine for Min Maha Giri Nat, the guardian of the household, with a coconut hanging on the front corner pole of the house.
>
> In addition to Min Maha Giri Nat, there are other spirits worshipped by locals for specific purposes. Farmers mainly pay homage to Ponmagyi Nat or Ponmagyi Shinma. This *nat* is worshipped on various occasions, from plowing to harvesting. At the Ponmagyi shrine, believers pray for a high yield of their crops, protection from pests, and favorable weather. The offerings include deep-fried red and white rice dough, a customary feast with cooked rice

and dishes. After the ritual, the offerings are shared among the worshippers and others nearby, promoting friendship and solidarity among the villagers as well as honoring the *nat*. On the third waxing day of Tabaung, people from all over the country pay homage to Ponmagyi Shinma. Although the name of the *nat* may vary depending on the local language, the practice remains uniform across the nation.

There is also a soapbox-sized shrine built on a single post at the edge of the farmland next to the river. This shrine is for Ywa-Taw-Shin or Ne-Daw-Shin Nat, known to have control over a specific area. They are often referred to as Bo Bo Gyi, A-Ba, or A-Bo. Worshipping these spirits is done to protect the land from erosion and destructive insects. Farmers working further up on the land invoke Ponmagyi and the same Bo Bo Gyis for the welfare of their farmland. In addition to these deities, there are also *nats* associated with parents and ancestors, known as ancestral *nats* or Mi-Saing-Pha-Saing *nats*. These ancestral *nats* may vary depending on the native region of each individual and their ancestors, who may have migrated from different parts of the country.

Migrant *Nats*

There are fishing people living in the area who pay tribute to their respective *nats*. At home, they pay homage to ancestral *nats*, while at the bow of their boats, they honor Ko Gyi Kyaw and the Taung Byone brothers. In the hull of their boats, A-Me-Ye-Yin is worshipped. The *nats* at the bow protect the people on the boat, while those in the hull ensure their bounty. Offerings for Ko Gyi Kyaw include a whole fried chicken and a bottle of liquor. For A-Me-Ye-Yin, offerings consist of pickled tea, incense sticks, and fried rice dough. The *nat* offerings can be as elaborate as the worshipper's wealth allows. As there are numerous *nats* to whom families pay respect, there can be thirty or more bunches of bananas and other related items. In addition to individual rituals, people celebrate a common *nat-pwe* every three years, praising almost all the *nats* nationwide.

One particular hazard in the river is the *nga-ywe* fish, a giant predatory catfish whose strength can tear down fishing nets. However, these fish are also regarded as respected *nat*-fish, and if caught accidentally, they are released back into the water. The Ayeyarwady dolphin is considered a friend to the fishermen, aiding their catch by driving schools of fish into the nets.

Except for the Taung Byone brothers, the *nats* mentioned below are not native to the areas. The reason for their presence in those areas is that people would travel to the Ayeyarwady valley to settle in fertile lands with freshwater livestock.

THE ORAL HISTORY of A-Me-Ye-Yin has been passed down through generations of believers. She was a sister of an underlord of Pontaung Ponya in the present-day Magway Region. She became a powerful *nat,* and travelers are very cautious when they have to go through the jungle in her jurisdiction. There are many possible original headquarter shrines, but based on the story, the one in the Magway Region is most likely the original shrine. Her influence is not limited to the rivers and livestock; inland communities in other parts of Burma also pay respect to A-Me-Ye-Yin to solicit her blessings.

THERE ARE FOUR *nats* known by the name Ko Gyi Kyaw. Although each Ko Gyi Kyaw has his own backstory, their behavior and attitudes are similar: they drink, gamble on rooster fights, and help others. The story of Pa Khan U Min Kyaw is particularly related to a part of the Chindwin River near the confluence of the Ayeyarwady. He was a prince assigned by King Min Khaung of Ava to dig a canal to irrigate water from the river into the moat of the newly built city of Pakhan. A fun-loving prince, he failed to supervise the canal project, which led to his execution by the king. However, Ko Gyi Kyaw's subjects were not happy with his demise as he had been a benevolent lord who pleased and protected them. He was later regarded as a *nat* worthy of worship, and people in Pakhan celebrate his festival every year.

THE HISTORY OF THE Taung Byone princes dates back to the eleventh century CE in the pagan period and is well known to the Burmese. One common version of their legend is that they were too busy playing marbles to heed the king's decree that all subjects should bring a brick to help build a great pagoda. For this act of lèse-majesté, they were executed. To commemorate and celebrate, three events are held on specific days.

The first commemorates the *nats* setting out to war and is held in the Burmese month of Nadaw, a Burmese lunar calendar month that usually falls in December. Another one is celebrated in the Burmese month of Tabaung (February or March), marking their return from the front line, when the villagers welcome them home. Neither, however, is as important as the festival that marks the beginning of the monsoon season, held at the brothers' principal shrine in Taung Byone, eleven miles to the north of the city of Mandalay. Hundreds of thousands of believers from all over Burma come to the village every year to pay respect to the *nats* at their shrine. The festival, believed to have been held since the eleventh century, is celebrated from the eighth waxing day of Wakhaung (July or August) to the full-moon day of the month. Each day of the festival has specific functions. On the fourth day, there is the Cho-Ye-Daw-Thone-Bwe, meaning the bathing of the *nats* in the Ayeyarwady River. In the past, when the water in the Shwetachaung Creek reached to the edge of the village, the statues of the *nats* were brought to their bathing spot along the village road, a unique part of the festival. Other *nat* festivals, following the example of the Taung Byone brothers, also perform similar rituals.

The Ayeyarwady Watershed

The Ayeyarwady watershed will occupy center stage for much of the coming analysis. Although a great deal can be said about watersheds in general, each particular watershed is individual, having a unique history, configuration, hydrology, and suite of riverine floral and faunal companions. An introduction to the Ayeyarwady watershed and its personality quirks seems to be in order.

A representation of this kind is not without its perils. An odd combination of a mug shot (a map), a deep historical biography, and an account of its seasonal moods, together with its assemblage of riverine creatures, risks misrepresenting the river as just so many static pipelines on the map, creating a fixed landscape and pumping volumes of water out to the sea. I shall endeavor to disturb any static understanding of the river and instead emphasize movement, which is never fully predictable.

Geological facts, though themselves in motion over tectonic time scales, are the setting to which rivers must, in the short term, adapt. Virtually all the mountain ranges in Southeast Asia are arrayed in a north-south direction. All the major rivers, as a consequence, are similarly oriented, with their headwaters in or near the Himalayan plateau and their estuaries in the Indian Ocean or the Andaman or South China Sea. The Ayeyarwady is no exception; it follows the valleys formed by tectonic action, particularly in the Sagaing Fault between Bhamo and Mandalay.

Although confined to the north-south basin, the Ayeyarwady has morphed radically in comparatively recent geological time.

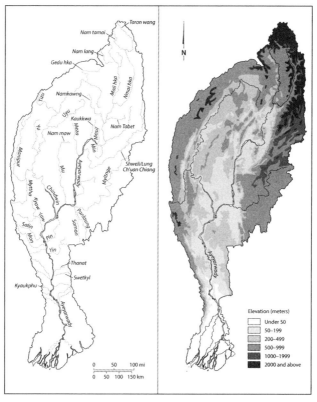

Map of major tributaries of the Ayeyarwady River (*left*) and elevation levels (*right*).

A mere fifty thousand years ago, it was part of a massive gulf that submerged most of what is now the Ayeyarwady and Sittaung basins, bisected longitudinally by what is now known as the Pegu Yoma range, itself largely submerged. The gulf is likely to have been re-created by the huge rise in sea levels beginning around eight thousand years ago, when pulses of glacial melt drove sharp increases in world sea levels. The silt-charged bed of the gulf accounts for the historic richness of its floodplain soils as well as contributing to the early productivity of the Dry Zone soils, which receive little in the way of rainfall today.

Many geologists believe that the paleo-Ayeyarwady occupied what is now the Sittaung's watershed to the east of the Pegu Yoma. Speculation about the cause of the westward shift of the Ayeyarwady to capture the Chindwin watershed centers around the volcanic activity in the vicinity of the Mount Popa-Wuntho arc roughly 5 million years ago.

For readers accustomed to the hydrology and ecology of rivers in temperate zones, the monsoonal and semitropical character of a watershed like the Ayeyarwady requires emphasis. The flood pulse, characteristic of virtually all rivers, is, in monsoon zones, far more massive and more predictable. Correspondingly, it is likely to create vaster floodplains and wetlands as it overtops its banks at peak flood stage. The effect of this surge (in the Ayeyarwady this surge is on average eight times its dry-season flow) is to create a far larger in-between landscape that is neither strictly terrestrial nor aquatic, but sequentially dry and wet to various degrees. Its flora and fauna reflect this ecology in ways that will be elaborated on below.

Tropical watercourses are distinctive.[2] Because there is no pronounced cold or dormant season, the central role played by leaf litter in the organic composition of temperate waters is absent. Aquatic plants and their detritus correspondingly take the place of leaf litter. Owing to the higher temperatures of the water in tropical streams, they support a greater density of bacteria, fungi, algae, and other forms of microbial life. As these basic foundations—along with insects that dominate the fauna of all rivers—of the food chain are richer, they are more biotically productive by an order of magnitude than temperate watercourses.[3] Due to the greater nutrient load and the lateral connectivity of the monsoon flood pulse creating a multitude of microhabitats, the species diversity in the river and in the riverine environment generally greatly exceeds that of temperate streams.

I will, as is conventional, briefly describe the watershed from its highland origins and tributaries to its eventual merger with the sea via several distributaries in the delta. The description will, less conventionally, be devoted to the river and its tributaries, not to the major human settlements along their banks.

The Ayeyarwady (ဧရာဝတီမြစ်), as opposed to its watershed, is said to begin at the confluence of the N'mai Kha and the Mali Kha (Myitsone မြစ် ဆုံ), which lies forty-two kilometers upstream of the Kachin State capital Myitkyina. Together, they drain roughly 11 percent of the entire watershed. The next substantial tributary to join the Ayeyarwady is the Shweli (ရွှေလီမြစ်), draining roughly 5.5 percent of the total and joining the Ayeyarwady from the east between Katha and Tagaung. It is known for irrigating extensive marshes near its confluence that are the habitat of endangered endemic species of duck and known for the lush vegetation (particularly rhododendron) of the floodplain. Moving downstream, the next major tributary (second only to the

Chindwin in the proportion of water drained, 11 percent) is the Myitnge (little river, မြစ်ငယ်, to distinguish it from the "big" Ayeyarwady itself). It too drains the watershed to the east of the Ayeyarwady and flows into it near Mandalay, at the site of the old capital at Amarapura.

The remaining two substantial tributaries, the Mu and the Chindwin, drain sections of the watershed to the west. The Mu River (မူး မြစ်) is by far the smaller of the two, draining less than 5 percent of the watershed owing to relatively dense settlement irrigation and, more recently, dams. More of its volume is consumed before it joins the Ayeyarwady than is the case with any other tributary. The upper, forested reaches of the Mu River are inhabited by the Kadu and Kanan minority peoples, who may well represent the remnant of the ancient Pyu who fled the Burman military colonists at the end of the first millennium CE.

The final and most dominant of all tributaries by far is the Chindwin River (ချင်းတွင်းမြစ်), which drains more than 27 percent of the watershed and contributes much of the sediment carried to the sea by the Ayeyarwady. Its total length of 1,200 kilometers is roughly half as long as the length conventionally ascribed to the Ayeyarwady itself.[4] During the high-water rainy season, from June to November, it is navigable by river steamer for more than 640 kilometers upstream from its confluence. The confluence itself is a vast, shifting zone of wetlands, aquatic diversity, and alluvial islands that appear and disappear as the river carves new channels at flood stage. After the massive pulse of water and sediment delivered by the Chindwin, the subsequent fourteen tributaries, confined in a greatly constricted watershed, collectively contribute little more than a third of what the Chindwin delivers.[5]

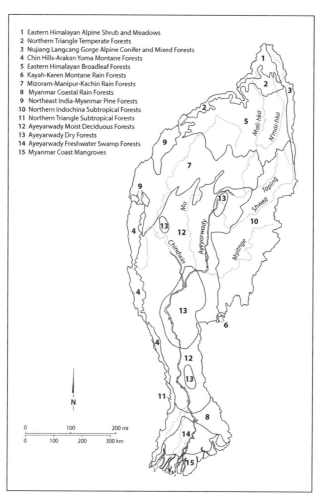

Map of ecological zones of the Ayeyarwady Basin.

Leaving the Dry Zone, the river passes Pyay, ancient site of the pre-Burman Pyu people (Sri Ksetra, first-century CE) who farmed the floodplain and built an early walled city there. Hinthada is the next notable settlement on the river. Downstream from Hinthada we are dealing with distributaries, not tributaries,

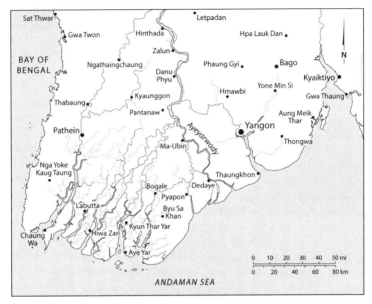

Map of the Ayeyarwady Delta.

fanning out across the delta. The most prominent of the distributaries that traverse the delta were formed over the last eight thousand years by the accumulation of sediment transported from the hills to the sea. The main distributaries are the Pathein River in the west, the Bogale in the center, and the Yangon River in the east, each bearing the name of the largest city that dominates its course.

A Slow River with a Vast Floodplain

The gradient of the Ayeyarwady is exceptionally low. Mandalay, the precolonial capital of the last Konbaung Dynasty monarch, is a full one thousand kilometers from the sea, but its elevation is a mere eighty meters above sea level. Thus, on average, it drops

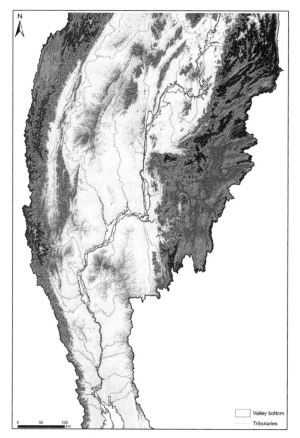

The Ayeyarwady floodplain.

only seven centimeters per kilometer on its way to the sea. Its descent is, of course, uneven, which means that there are large portions of its course that are nearly flat. Such is particularly the case with the Dry Zone south of the pottery-making town of Kyaukmyaung. Between Mandalay and Magwe, located more than three hundred kilometers downstream near the southern extremity of the Dry Zone, the river only loses twenty-one meters in elevation. Hinthada, conventionally seen as the head of the

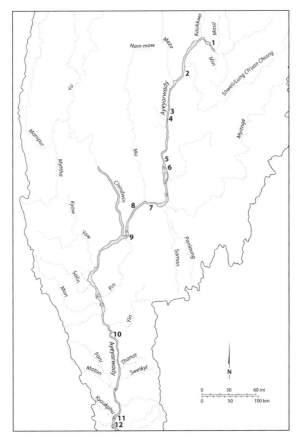

Location of several wetland areas in the Ayeyarwady floodplain.

delta, is only seventeen meters above sea level, although it still has more than two hundred kilometers to travel before it reaches the sea. One major consequence is that the lower half of the delta is a tidal zone, far more affected by exceptional tides and storm surges, while the upper delta is under the influence of the freshwater flood pulse.

The consequence of a large floodplain and a low gradient is an erratic, meandering channel-shifting river that, along with abundant monsoons surges of rainfall, makes for a floodplain with periodically flushed, vast wetlands.[6] Both floodplains and associated wetlands were the effects of millennia during which the river was largely unaffected by human activity.

Seasonality

The Ayeyarwady is not the same river from millennium to millennium, century to century, year to year, or, for that matter, day to day. Thanks to the monsoon, which accounts for 92 percent of the total annual rainfall in Burma, the volume of water the river carries at high water in September is, on average, eight times the average water volume at its lowest in February. A river that has swollen eightfold can hardly be understood to be the same natural object as its tranquil and diminutive predecessor earlier in the year.[7] Not only is it wider and higher, it also abruptly seizes floodplains, riverine forests, wetlands and marshes, and nearby flatlands that are, for most of the year, arid. In the occasional strong flood pulse year, the Ayeyarwady extends its reach to rarely inundated areas, transforming their flora and widening their biodiversity. The river at flood stage re-creates the in-between: that considerable landscape that is neither terrestrial nor aquatic but amphibious and therefore exceptionally diverse.

Our sense of rivers is that they flow inexorably downstream. There are, however, exceptions. The most striking example along the Ayeyarwady is at its confluence with the Chindwin at the apex of a big monsoon flood. Owing to the exceptionally low

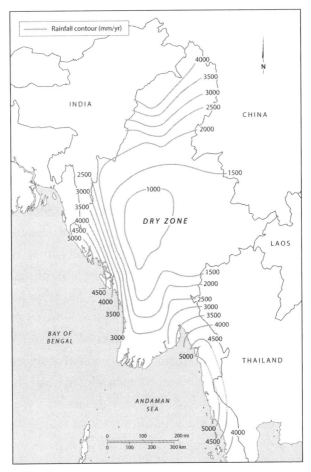

Rainfall pattern in Burma.

gradient of the Dry Zone upstream from Sagaing, there is occasionally a ponding effect all the way to Mandalay, as the downstream flow is blocked and temporarily *reversed*.[8]

More generally, on an annual basis the flood pulse moves north to south, following the river's course. The timing and magnitude of the pulse's downward trajectory can be roughly calibrated from the pattern of rainfall.[9] Where the river encounters a flat landscape, it spreads out laterally, thereby slowing its downstream momentum until it encounters natural barriers that restrict its lateral movement. As the river moves, so too do virtually all the life-forms dependent on the riverine environment. Fish migrate to newly available sites for spawning and nutrition. Dormant seeds and plants dependent on a flush of water sprout and spread. The inundation of in-between landscapes temporarily drowns many terrestrial plants and creates a favorable setting for aquatic plants. Insect populations are transformed, depending on their adaptability to the shifting ecosystem. Because fish, insects, and plants are repositioning themselves, so too are their predators: the birds, mammals, amphibians, and reptiles whose movements are synchronized with those of their main sources

Illustration of the Chindwin confluence after the rainy season (*left*) and during the dry season (*right*).

of nutrition.[10] The first mover, the very premise of this massive rearrangement of the living riverine world, is the periodic clockwork of the river's flood pulse.

As a riverine species, *Homo sapiens* is no exception to the adaptations set in motion by the river. We have already described the dependence of hunters and gatherers and most early civilizations on the density of resources afforded by proximity to the river. For sedentary agrarian communities, the development of flood-retreat cultivation represented the only form of grain cultivation that was competitive with hunting and gathering in terms of the expenditure of labor for the return in nutrition. In fact, the flood pulse itself made possible a form of fishing that was comparable to flood-retreat agriculture. One might call it "flood-retreat fish capture." Its practice almost certainly predated agriculture and it is still, along with flood-retreat cultivation, practiced along the Ayeyarwady today. The logic is simple. At a manageable and relatively narrow confluence during the high-water monsoon, the fishers drive stakes (usually bamboo) into the streambed close enough together to prevent most fish from passing downstream but, at the same time, allowing the water to pass through. The trapped fish are then easily harvested when the water level subsides, often by a small door that funnels them in small lots into baskets and nets. In some cases, moveable weirs are constructed and maneuvered downstream by hand to drive fish into depressions of smaller upstream feeder streams in the capture zone.[11] Here the human techniques of fishing mimic those of other predators except for their (slight) modification of the waterscape. As we shall see, the modern era is marked by a massive scale of modification, a point at which differences in quantity amount to a qualitative revolution.

The Sediment Pulse

The flood pulse, understood as seasonal high water, is not the only pulse animating the river. Just as important is the pulse of sediment carried by the flood pulse at its peak. The river current is a conveyor belt carrying silt, clay, sand, gravel, pebbles, and even boulders downstream. So consequential over time was this conveyor belt for the Ayeyarwady that it has created most of the Ayeyarwady Delta south of Hinthada, the most productive agricultural region of modern Burma. This hydraulic land-making feat was accomplished in a mere—geologically speaking—seven or eight thousand years. River towns that were near or at the coast two thousand years ago are today inland river ports. The delta itself is both a natural and an anthropogenic landscape—natural because of a relatively shallow coastal shelf and a load of sediment from the friable terrain of its major tributary, the Chindwin, and anthropogenic because of human deforestation and drainage from at least the tenth century forward.[12] Prior to the advent of major dams, the Ayeyarwady was consistently ranked among the five most sediment-laden rivers on the planet.

The sediment pulse is likely to coincide with the flood pulse when the volume and speed of the water is sufficient to move heavier material. There is an obvious physical and gravitational logic to the sediment pulse, depending on how much force is required to transport it. The finer silts and clays are most easily dislodged from the riverbank and bed and suspended in the water column while, obviously, sand, gravel, pebbles, and cobbles—not to mention boulders—require far more force to move. Lighter materials are swept up first and heavier materials later as the flood pulse strengthens; conversely, as the flood pulse weakens, the heavier materials are the first to precipitate out and fall to the riverbed. In this sense, *there is no single sediment pulse,* but rather a range of distinct pulses depending on the kind of sediment being moved.

The sediment pulses are as consequential as the flood pulse for the movement and shape of the river over time. As the flood pulse weakens, particularly in flat terrain, the deposition of accumulated sediment builds up to the point of forming a bottleneck that obstructs the preexisting channel. The river is forced out of its banks, spreading over the adjacent plain while seeking a new channel to the sea. It is this effect of the sediment pulse, as much as the flood pulse alone, that is responsible for what is termed channel migration. Much of this channel migration is on a Lilliputian scale, causing micro changes in channels, but some large-scale blockages have provoked radical changes in the course of the river. We have earlier described the changes that characterized the movement of the Yellow River as it swung north and then south of the Shandong Peninsula. The Ayeyarwady itself was once thought to have followed the bed of the Sittaung River before assuming its present course to the west of the Pegu Yoma range and capturing the Chindwin.[13]

The biological productivity and diversity of the riverine landscape are the joint production of the peak flood pulse and the interruptions made possible by sediment blockage. These processes send the river on a new journey to previous channels and floodplains or sculpt new ones, refreshing old wetlands and creating new ones, disturbing existing habitats and making new ones. To the degree that channel migration is the engine of watershed abundance, that abundance is the creature of the deposit of sediment.

Alluvial Islands

Nothing exemplifies the theme of river movement more than the constant formation of alluvial islands. They suddenly appear, disappear, grow, erode, or attach themselves to the bank only, perhaps, to be cut off again by the next monsoon or storm surge. They are, of course, a product of flood sediment, as is the entire delta of the Ayeyarwady, a vast peninsula built by the flood pulse conveyor belt. Alluvial islands are most common in the stretches of the river where the gradient is low and pulses of sediment from upstream are common. They are found as far upstream as Bhamo, near the Chinese border. Several of the largest lie immediately

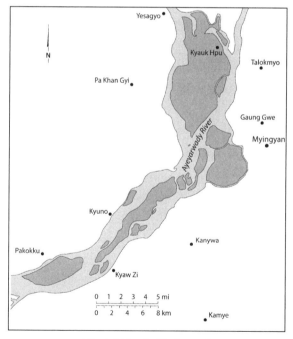

Alluvial islands (dark gray).

downstream of the confluence of the Chindwin and the Ayeyarwady, extending all the way to Pakokku and beyond.

The largest number of alluvial islands, as one might expect, are to be found in the delta where the low gradient and the cumulative sediment load carried by the peak monsoon floodwaters combine to remake the landscape annually.

To judge from the deep history of movement depicted in Fisk's maps of the Mississippi meander zone, it is reasonable to suppose that much of the riverine population of birds, reptiles, amphibians, and mammals—including *Homo sapiens*—now dwell on land that was once an alluvial island. By the same token, given the vagaries of river movement, these species may, in future and without budging an inch, find themselves again on an alluvial island.

As habitats, alluvial islands are cherished by humans who live on or near the river. Benoit Ivars, an expert on the Ayeyarwady's alluvial islands, judges that they are anywhere from three to ten times more productive than ordinary padi land on the adjacent floodplain. The desirability of such fresh floodplain land is captured in the fact that it is often referred to as golden or silver, a gem or a treasure. For an alluvial island that is growing, there is a natural floral succession. Kaing, a pioneer grass, is the first colonizer and serves to bind the loose, sandy soil. If subsequent monsoon silt deposits raise the alluvial plateau above the average flood peak, more of the soil becomes suitable for clearing, usually for the cultivation of padi rice or other moisture-tolerant crops. Ideally, for *Homo sapiens,* at least, parts of the island may rise sufficiently to encourage human colonists to plant orchard crops in addition to padi and to establish villages. This is most likely when the currents of the river have shifted to become less immediately threatening. Even then, the alluvial island is not immune to the river's moods. Local wisdom, quoted by U Soe, a poet in the village Ivars studied, held that "an island lives for [only] 100 years."[14]

The movement and in-betweenness of alluvial islands confound the assumed certainties of terrestrial rule and statecraft. Who may claim the new resource and on what basis? Once a potentially valuable alluvial island emerges, it becomes the object of what amounts to a golden-land rush by competing nearby villages. The typical first step is to create facts on the (new) ground by clearing brush and staking out the dimensions of a potential padi field, under the apparent assumption that "possession is nine-tenths of the law." Frequently, there were—and are—two competing villages on opposite sides of the river asserting a claim to the new land. A common-law practice, probably predating state-certified land titling, was used to resolve such disputes without resort to force. A buoyant object, usually wood, was placed in the river current well upstream of the new alluvial island. As it floated downstream, driven by the prevailing current, it would presumably pass the island on the side where the current was strongest. That channel was deemed the river and the village separated from the island by the river would cede its claim to the contending village now separated from the island by a minor or branch channel. The accompanying figure depicts the results of a common-law procedure along these lines. The black tire channel on the map, reconstructed from U Soe's account, refers to the black tire used to identify the main channel and thereby resolve the dispute.[15]

The flood- and sediment-created new habitat is desired not only by humans but also by pioneer flora, fish, bivalves, birds,

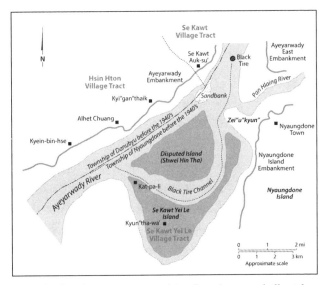

Method to determine ownership of newly created alluvial islands (dark gray). Note the path of the "black tire channel," which indicates the change in ownership of the disputed island.

and insects adapted to its environment. Only humans, however, can boast of rulers and states that claim the right to allocate legal title to whomever they designate. Only humans also endeavor—using bamboo pilings, stones, sandbags, and small weirs—to freeze or extend their new property and prevent the river from reclaiming it. As U Soe's adage has it, their victories are inevitably temporary.

FOUR

Intervention

> Total control for greater Wealth.
> —U.S. BUREAU OF LAND RECLAMATION, 1930S

Preindustrial humans, as we have seen, had a massive cumulative impact on watersheds wherever fixed-field agriculture was practiced. The magnitude of that impact was proportionate to their numbers, the amount of land they had cleared for crops, and the quantity of fuelwood necessary to create their proto-industrial products: for example, pottery, bricks, plaster, and early metals. Compared to the age of industrial reengineering of the riverscape that followed it, this period bore at least three distinctive features. First, it was geographically concentrated on areas of relatively high population density. Estimates of the world's population in 1400 CE are less than four hundred million, a mere half of a percent of today's population, and a good many of these people were hunters and gatherers. Second, the tools at their disposal—above all, fire, the waterwheel, draft animals, and early iron implements—circumscribed their impact. Finally, humans' impact on the riverscape was largely an inadvertent by-product of their subsistence practices rather than a deliberate attempt to relandscape the river. The distinction, of course, is of little comfort to the fish, birds, mammals, plants, and trees whose lifeworld was in jeopardy.

The watersheds into which preindustrial cultivators intruded were *not* pristine, undisturbed, primeval landscapes. All creatures in the watershed are landscapers and niche-builders. They ceaselessly remake their habitat, often in competition with other living creatures. The ecological mosaic that all these creatures—fish, birds, mammals, insects, amphibians, plants, and soil and aquatic biota—create, together with the sheer physics of the landforms and hydrology, might aptly be termed the vernacular riverscape: the vector sum of the landscaping forces at work. Our attention is drawn to the beaver and *Homo sapiens,* both of which have had an outsized historical role in sculpting the watershed, but they are only two of a myriad of species reworking the riverscape. Only in the full-blown Anthropocene does *Homo sapiens,* organized into states and empires, come to exercise hegemonic power over *engineering* the landscape.

The onset of the full-blown Anthropocene and the twilight of the vernacular riverscape in the nineteenth century were marked by two world-altering events: the industrial revolution and states determined to refashion nature for the benefit of a single species. Industrialization, including the fossil fuel revolution, exponentially expanded human power to intervene in nature. Taken together, dynamite, earth-moving machinery, and reinforced concrete made it possible to reengineer the riverscape on a massive scale. Much of this newfound power was based on the internal-combustion engine—hence the appropriateness of the term *engin*eering—perhaps the culmination of the earliest quest of our species to domesticate fire and make it do our bidding. If, in earlier epochs, we had been forced to move *with* the river, it was now within the human grasp to *move the river*. Though the new mastery of the river was hardly as comprehensive as engineering utopias promised, there is no denying the heady sense of power felt by humans who had been playthings of a capricious watershed and now contemplated becoming river masters. Instead of cursing the unpredictable, erratic river, with landscaping tools in hand, we could now, it seemed, design a river to serve the purposes we assigned it. Mastery over nature!

Defining the purpose(s) to which a river might be put required a structure of power that would both decide the purpose in question and assemble the resources and plans necessary for its execution. That institution was, of course, the empire or the modern nation-state. Two coterminous developments fueled the enthusiasm for large-scale landscaping. First was the rise of engineering and hydrology as professional specialties in the context of high-modernist achievements in machinery, skyscrapers, aircraft, manufacturing, and industrial chemistry. Ambitions to redesign nature for human purposes had never seemed so realistic.[1] The second factor was the nearly universal sentiment, pioneered by Bismarck, that the purpose of the state, be it autocratic or quasi-democratic, was to improve the welfare of its population.[2]

Two objectives dominate this chapter. The first is a description of the most prominent industrial-scale transformations of the Ayeyarwady River over the past century and a half. Virtually all of them merit the designation of full-blown Anthropocene effects, as they were made possible by modern technology. The second objective, more interpretive in nature, is to understand the broader meaning and implications of human and state efforts to reengineer the complex hydrological and biotic processes we call a "river" for the purposes of a single species. Understandably, the logic of such interventions has much to teach us about the relation of modern humans to nature, broadly understood.

Declining Fish Catch: Illegal Fishing and Pollution

Related topics that have dominated local conversations over the past few decades are the decline of fish populations, the virtual disappearance of some desired species, and a catch dominated by smaller and more juvenile fish. Once again, Maung Maung Oo and Naing Tun Lin interviewed an array of fishing folk to learn their local wisdom on fishing techniques, pollution, and the widespread decline of the catch. Their narratives have been edited for clarity and concision. The first stories were collected by Naing Tun Lin.

> THE TOWNSHIP of Htigyaing is renowned for its bountiful fish production, surpassing even that of Inn-Ywa. Both towns were established in the eleventh century CE as fortress towns, erected to repel invaders. They've grown to become more substantial than neighboring settlements.
>
> According to Ko Tun Aung, a Htigyaing native, the abundance of fish was astonishing until the 1980s. Anyone armed with a hand fishing net, known locally as *yet-the,* could effortlessly catch four pounds of fish within minutes. The region's uneven terrain is punctuated by troughs filled with floodwater or rain during the monsoon season. These troughs, disconnected from the river when the rainy season ends, transform

into sanctuaries for fish. This is attributed to the presence of *Hsou-Ka-La Chone,* a type of vegetation that provides refuge for the fish. The fish were caught using large bamboo wicker traps that prevented their return to the river as the water receded.

Fishermen used to dig a five-foot-deep pit measuring ten by ten feet to store their catch. As the fish swam over the traps, they were hand-caught and placed in the pit. When the pit became full, a brass gong hanging nearby was struck to alert villagers to gather to share the catch. Some of the fish were sold, while others were salted in the pit for future use. Certain fish were pickled and distributed to other parts of the country. However, locals tended to avoid fish without scales, opting for other choices. The most highly regarded fish was the *Nga-Thine,* although it has become rare due to human activities that have harmed freshwater life.

After the late 1980s, the fish population in freshwater bodies began to dwindle. As people migrated to the area and initiated activities detrimental to the fish population and their habitats, the decline accelerated. The destruction of flora that provided shelter for freshwater led to a dramatic reduction in fish and other aquatic species. Timber production, particularly the extraction of valuable teak and hardwood trees from the monsoon forests, began to proliferate. The practice of placer mining, involving small-scale gold production through panning, and later large-scale gold production by dredging sand from the riverbed further disrupted the ecosystem's balance.

Fish breeding rates also plummeted during this period. As a survival strategy, the fish began to migrate upstream. After 1988, they gradually moved to Bahmo, becoming scarce in

Htigyaing-Mya Daung areas. This migration was driven by the need to find habitats with fewer dangers to their survival.

With the decline of the fishing economy, which was deeply intertwined with the local community's way of life, the rites and rituals common among the locals began to wane. For example, around 1988, near the town of Htigyaing, a plain by the river was leveled for government timber production and transportation. Here, a deceased Ayeyarwady dolphin was discovered stranded in the logging bay. The locals, adhering to a tradition that forbade them to kill dolphins in the river, buried the carcass in the ground. The elders in the community strongly believed that butchering the dolphin for oil would bring bad luck. Sadly, their wishes were not honored, and the dead dolphin was cut into pieces the following day. Although the consequences of their actions remain unknown, the entire logging bay was later eroded by the river when the rains arrived.

KO KYAW SAN, now fifty-four, had been fishing in the Ayeyarwady River since the age of seven, accompanying his father in a small boat. When he grew older, he was strong enough to walk on the bank manually dragging the boat upstream against the current, a common practice until the mid-1990s. They would make periodic stops, strategically placing their fishing nets (one hundred to three hundred feet long) attached to poles on the river's shallow bed. The locations were marked for reference, and the journey would continue.

A customary ritual among the river's fishermen is offering their meals to the local *nats* before eating themselves. This practice seeks their blessing for safe and fruitful fishing as well as protection from potential dangers. After lunch and a brief rest, they would cast their fishing nets again in the afternoon until dusk. At night, they would find a suitable spot on a sandbank near their fishing net to sleep. Notably, they avoided sleeping near farmlands.

Prior to settling down for the night, they would draw a circle on the ground and once again pay their respects to the local *nats*, asking for permission to stay. This ritual ensures they feel secure.

The entire fishing process might span more than a week, depending on their catch. Once they collect their catch, it is sold at the market, and they may move to another location to set their nets again. They continue until they've earned enough to support their households.

Maung Maung Oo's input starts from here.

UNTIL 1988, fishermen adhered to unwritten rules and traditions to preserve the river's fish stocks. They refrained from fishing during spawning and breeding seasons, allowing fish to grow to a larger size. Their strong moral code dictated their behavior, including ostracizing those who did not follow these practices. They believed that following these customs was not only ethical, it also safeguarded them from being labeled illegal fishers.

Fishing methods varied depending on the target fish species in the river. Smaller fish, for instance, were released if they were unintentionally caught in nets.

For example, long-whiskered catfish, known as *Nga-gyaung* (*Sperata aor*), were baited with live prawns on hooks attached to a line. The line, connected to stones on both ends, was sunk into the riverbed. The next day, the fishermen would retrieve the line using a hooked rope from their boat. The same method applied to catching *Nga-yaungs* (*Cephaloocassis jatia*), but with a different bait—a piece of locally produced *Shwe-wah* soap.

To attract *Nga-myins* (white catfish, or *Silonia silondia*), moths from nearby lakes were used as bait. Since these fish reside in shallow waters, the line was fastened to bamboo poles positioned in the shallows. *Nga-yan-goungdos* (walking snakehead, or *Channa orientalis*) were caught with live crickets on hooks, aligned along a line attached to bamboo poles resting above the water's surface. Snakeheads, upon seeing their prey above the water, would leap to seize the bait. As snakeheads are powerful and agile, they could dislodge the bamboo poles. Fishermen had to react quickly to catch them before they escaped.

There were also freshwater stingrays (*Urogymnus polylepis*) in the Ayeyarwady, but they are now nearly extinct. In the past, these creatures were lured using heated iron chains, as the smell of iron from the furnace attracted them. Bait types varied according to the fishermen's preferences.

Fishermen did not fish in the river year-round. They adapted their fishing methods based on the river's seasonal flooding and receding. When the river began to rise in late April, they placed bamboo traps in the flood-retreat farm areas at the water's edge. As the monsoon season began and the river flooded the fields, they laid out nets and fishing lines. During the peak of the rainy season from July to September, they took a break from fishing. They resumed their activities in October, following the spawning and breeding seasons. Fishermen were categorized by the type of fish they caught—some focused on larger fish, others used cast nets for medium-sized fish, and some employed gillnets for surface-dwelling fish.

These traditions were vital in maintaining the ecological balance of the river and ensuring a healthy fish population. Until economic hardships and political instability took a toll on the nation, these customs, though sometimes considered archaic, preserved what we are now losing.

The following is Naing Tun Lin's account.

KO KYAW SAN, a knowledgeable fisherman who lives on the island of Than Bo, right in middle of the Ayeyarwady to the west of Mandalay, feels that the current decline in the fish catch, which is barely 20 percent of the yield in the 1980s, is mainly due to changes in the way they fish the river. To him, as a veteran fisher, the fisherfolk today are far more reckless and cruel. Today's fishermen see the fish stock in the river far differently than Ko Kyaw San and his fellow fishing people did. There are many reasons behind their attitude toward their business, but today's perspective is fundamentally determined by scarcity of fish. When the fishermen have to rely on fishing only, they strive to have the maximum catch by any means possible.

Here begin the findings of Ko Maung Maung Oo.

THE METHODS INCLUDE, but are not limited to, electric-shocking, using apparatus that is beyond what has been approved by the governments, poisoning from upstream, and using explosives in the creeks licensed for fishing. They are atrocities committed by fishing people out of desperation.

Locals see other major causes for the declining fish population in the river. Sewage and industrial wastes discarded directly into the river in Mandalay, Bhamo, and other cities along the Ayeyarwady are a major problem. (The industrial wastes discarded from industries in the Maungkone area near Tagaung have devastating effects on the river and its flooded areas.) They released their industrial wastes into the creeks and channels, which are connected to the Ayeyarwady River, to look as though they were avoiding polluting the big river. Gold mining along the river is the main reason for the chemical pollution. As gold is extracted by using mercury and cyanide, the chemical residues are usually piled up near the river. The river gets polluted during monsoon season when there is a sudden downpour, pushing the waste down into the river. There are other effects, such as the shift in the course of the stream caused by tailings disposed of by mining vessels. This results in sediment formation, as the silt is formed not only of alluvia but the coarse sand that is dredged out of the riverbed, making flood-retreat cultivation impossible for farmers who work the exposed islands. All this has had a huge impact on the habitats of the aquatic lives in the river.

Inland water ports are also one of the problems posing a threat to locals. Regardless of the size of the boat, the fuel oil is often accidentally spilled into the river at the port, polluting both the bank and the water.

The Shwekyetyet jetty near the two Ayeyarwady bridges connecting Mandalay and Sagaing has become a major port for transporting construction materials, fuel oil, coal, logs, and other high-volume commodities. Since 2004, the activities at this location have been severely damaging to the surrounding riverine area. When oils from tankers are transferred at the jetty, there is a considerable amount of leakage into the water and the ground. Child scavengers dip sponges fixed onto sticks into the spilled oil and squeeze those sponges into buckets of diluted oil to sell the result. There are no fish here, and the water smells so heavily of fuel oils that people can no longer bathe or launder clothes.

A lesser-known effect of river traffic is the damage caused by the wake of the waves they generate. There are rules and regulations regarding vessel speeds, but they are widely ignored. Many operators of the vessels are openly disdainful, traveling at speeds that cause erosion of the river bank. In one village named Ponna-Chan (the ranch of the Brahmin) in the township of Tada-U, the bank has already been taken down by the wake created by those boats.

Industrial-Scale Impacts on the Ayeyarwady

Before detailing the massive interventions made possible by industrial-scale technology, it bears repeating that the cumulative effects of demographic expansion, land clearing, and drainage for agriculture and fuel detailed in the previous chapter continue to operate in the full-blown Anthropocene. In fact, their cumulative impacts are vastly amplified by the population explosion that characterized the late nineteenth and twentieth centuries. Quite apart from the new technological tools at their disposal, in colonies and independent states seeking to reengineer the river, the growth in population alone would have massively intensified the clearing of forests for cropland and fuel that had already transformed the riverine landscape in the preceding thin Anthropocene. Deforestation proceeded apace, now accelerated by industrial methods. The map shown here, which covers only the fourteen years between 2000 and 2014, is historically deceptive because areas of the Dry Zone that were deforested much earlier show little further deforestation, which is concentrated along the upper tributaries of the watershed and the mangrove forests of the lower delta.

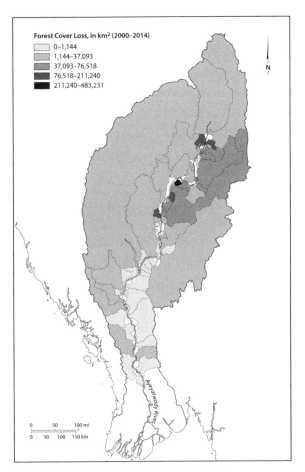

Total forest loss, 2000–2014.

Population figures for this earlier period are indicative of the pressure to clear forests and drain wetlands. Contemporary estimates of the current population of Burma hover around 55 million.[3] This figure is a *fivefold* increase from a plausible estimate of the country's population of 10.3 million in 1900. In keeping with the population growth of much of the Global South, the population of Burma has more than doubled since 1960. Whether or not the population boom merits classification as an Anthropocene effect is worthy of debate.

What is not debatable, however, is its consequences for the riverine landscape. Today, the Ayeyarwady Basin's remaining floodplain wetlands are only a quarter to a third of their estimated extent at the time of independence in 1948.[4] One can hardly overestimate the catastrophic ecological consequences of this loss of seasonal habitats for species diversity and abundance. The loss severely compromises the lateral connectivity—the lungs of the river's life-forms—made possible by the flood pulse. Much of the habitat loss took place in recent decades. Twenty percent of total basin forest cover, it is estimated, was lost in the two decades between 1990 and 2010, and one can reliably suppose that the loss has only accelerated since then. Much of the loss was concentrated in the delta and along the tributary watersheds of the Chindwin, Mu, Myitnge, and Shweli Rivers. Fully 20 percent of the mangrove forests in the lower delta were felled in a single decade, from 1995 to 2005.

The accompanying table depicts the forest types to be found in the basin and the degree to which they are critically endangered.

The pattern of forest loss closely tracks the historical concentrations of population and fixed-field agriculture (the Dry Zone). The moist, deciduous forests and the freshwater swamp forests

FOREST TYPES

Eco code	Eco name	WWF name	Status	Percentage of the area
PA1003	Eastern Himalayan alpine shrub and meadows	Montane grasslands and shrublands	Relatively stable/intact	1.5
IM0402	Northern Triangle temperate forests	Temperate broadleaf and mixed forests	Relatively stable/intact	2.6
PA0516	Nujiang Langcang Gorge alpine conifer and mixed forests	Temperate conifer forests	Critical/ endangered	2.1
IM0109	Chin Hills–Arakan Yoma montane forests	Tropical and subtropical moist broadleaf forests	Critical/ endangered	4
IM0401	Eastern Himalayan broadleaf forests	Temperate broadleaf and mixed forests	Relatively stable/intact	0.1
IM0119	Kayah-Karen montane rainforests	Tropical and subtropical moist broadleaf forests	Relatively stable/intact	0.1
IM0131	Mizoram-Manipur-Kachin rainforests	Tropical and subtropical moist broadleaf forests	Vulnerable	14.7
IM0132	Myanmar coastal rainforests	Tropical and subtropical moist broadleaf forests	Vulnerable	3.6
IM0303	Northeast India–Myanmar pine forests	Tropical and subtropical coniferous forests	Critical/ endangered	2

Eco code	Eco name	WWF name	Status	Percentage of the area
IM0137	Northern Indochina subtropical forests	Tropical and subtropical moist broadleaf forests	Vulnerable	15.5
IM0140	Northern Triangle subtropical forests	Tropical and subtropical moist broadleaf forests	Relatively stable/intact	12.9
IM0117	Ayeyarwady moist deciduous forests	Tropical and subtropical moist broadleaf forests	Vulnerable	27.6
IM0205	Ayeyarwady dry forests	Tropical and subtropical dry broadleaf forests	Critical/ endangered	7.7
IM0116	Ayeyarwady freshwater swamp forests	Tropical and subtropical moist broadleaf forests	Critical/ endangered	3.5
IM1404	Myanmar Coast mangroves	Mangroves	Critical/ endangered	2.4

Source: Charles-Robin Gruel, Jean-Paul Bravard, and Yanni Gunnell, *Geomorphology of the Ayeyarwady River, Myanmar: A Survey Based on Rapid Assessment Methods* (Washington, DC: World Wildlife Fund, 2016), table 3, pp. 17–18.

that, together with the Dry Zone, represent roughly 40 percent of the basin's riverine forests, have been extinguished or are endangered by human engineering efforts to prevent the river from claiming its annual floodplain or by humans extracting water that once nourished forests to irrigate dry-season crops on drained lands. Embankments and dikes—twenty-one hundred kilometers of them in the delta alone—and dams with reservoirs designed for irrigation have penned in the river and contributed to the elimination of seasonal wetlands and the habitat they provide to riverine and aquatic life. The result is a Great Drying, accelerated by industrial technology. A map showing the incidence of pumped irrigation both from watercourses and from dams on tributaries with reservoirs designed for dry-season crops is a graphic indica-

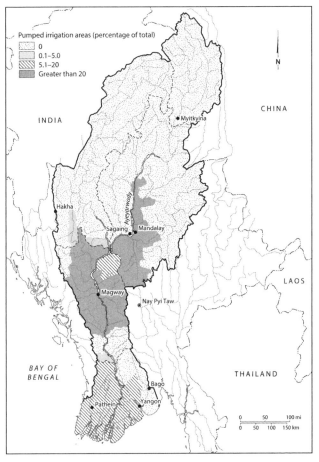

Pumped irrigation areas in the Ayeyarwady Basin.

tion of how industrial technology and impounded water combine to redistribute water for human purposes.

Hydroelectric dams are a much later and far more capital-intensive endeavor requiring masses of reinforced concrete and are sponsored, if not financed, by states. They represent, virtually without exception, a radical severing of the river's vertical connectivity. Such dams have been constructed on numerous tributaries (for example, the Mu, Chindwin, Myitnge, Shweli, Malikha, and Nam Kha), but not on the Ayreyarwady main stem itself.[5] The effects of such dams on the hydrology of the river and on riverine flora and fauna are massive. They block the transport of nutritious sediment to the downriver floodplains; they decrease the peak monsoon flow and thus the habitat reach of floodwaters into adjacent seasonal wetlands; they create backwater reservoirs that are biotically poorer than the flowing river they replace; and they block the passage of all migratory fish to upstream spawning sites. In an unpenned river, migratory fish transport nutrition upstream. Once that passage is blocked, it follows that there is correspondingly less nutrition available to sustain the diversity and abundance of birds, waterfowl, amphibians, reptiles, and mammals that depend on this.

Large dams may be the most modern form of imprisoning a river, but in terms of their impact on its hydrology, they are rivaled by the effects of levees, dikes, and embankments designed to prevent the river from occupying its floodplain. Unlike hydroelectric dams, dikes and levees are not an innovation of the industrial era. They are as old as the dawn of sedentary agriculture as farmers attempted to protect their crops and dwellings from monsoon floods and storm surges. What is distinctive about industrial-era dikes and levees is their relatively greater height

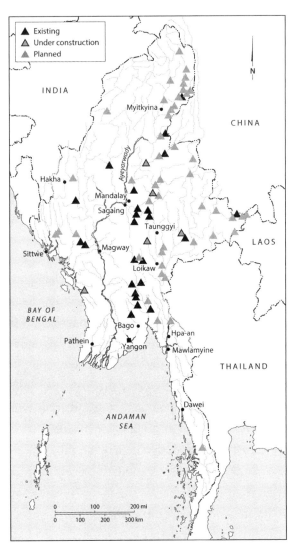

Location of hydropower dams in the Ayeyarwady Basin.

and extent made possible by earth-moving machinery and state financing far beyond what cooperative village labor alone could accomplish. There is not, to my knowledge, a basinwide map of embankments and levees because exceptional floods and typhoons, such as Nargis in 2008, periodically sweep away existing dikes.[6] The closest one can come to the extent of embankments—

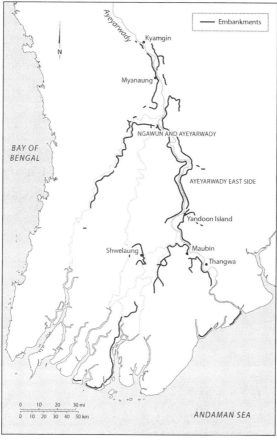

Embankments in the Ayeyarwady Delta region.

and this in the delta region alone—is an out-of-date map adapted from an assessment of the watershed in 2017.

Many of the embankments in the delta are constructed in a horseshoe pattern to divert monsoon floodwaters to the margins of the area of settlement and cropland. In the short run, it is an effective strategy. In the long run, like many flood-prevention measures, as we shall see, it generates dangers of its own. When waters and sediment are diverted to the exterior of the horseshoe, the deposition of sediment raises the land at the exterior well above the interior lands sheltered by the ramparts. The result, in the event of a major flood, is that the backed-up floodwaters pour back into the lower interior within the horseshoe, destroying both crops and dwellings.

We have thus far concentrated on the imprisoning of the river's water: both vertically, in the case of hydroelectric dams, and laterally, in the case of embankments and levees. But what of the waters thus penned in? What are the industrial effects on these waters?

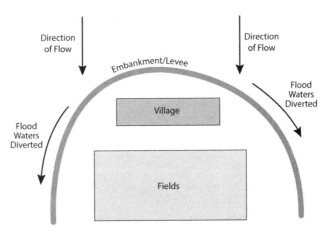

Model of village, cropland, levees, and delta.

The short answer is that the pollution of the river has grown exponentially in the industrial era, threatening or outright eliminating the habitat necessary for aquatic life in the watershed and along its banks. The major source of contamination comes from mining: gold, copper, tungsten, lead, zinc, silver, nickel, and, most recently, prodigious amounts of sand and gravel. Gold mining on an artisanal scale long predated the industrial revolution and continues apace today, being responsible for the greater part of mercury (used for millennia to extract gold from the ore) contamination. What has changed is the application of industrial technology, which has vastly magnified the effects on the watershed. Artisanal mining is now overshadowed by dredges, motorized barges equipped with hydraulic pumps capable of dislodging and ingesting vast amounts of ore-bearing soils from the riverbanks and riverbed to extract gold. The activity is ubiquitous throughout the watershed, destroying plant life and leaving a barren wasteland of tailings as well as mercury and sodium

Gold mining.

cyanide pollution. Since artisanal gold mining requires very little in terms of capital, it has been and remains a subsistence resort for poor landless Burmese under military rule.

Burma is a quite unique amalgam of centuries-old traditions of preindustrial mining—gold, jade, sapphires, and rubies—on the one hand and a rich portfolio of minerals indelibly linked to the industrial world—for example, natural gas, oil, tungsten, antimony, chrome, and rare earth metals. The revenue that can

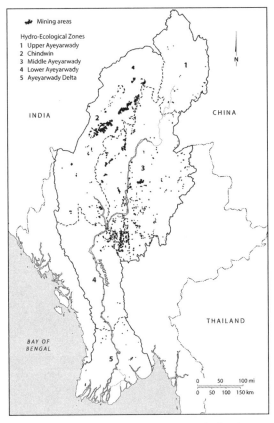

Mining activity in the Ayeyarwady and Chindwin watersheds.

be squeezed from controlling such mining has become a hotly contested prize between the military regime on the one hand and ethnic armed groups and their armed civilian Burman allies seeking to oppose and unseat the military junta. One result, comparable to the exploitation of the teak forests, has been unregulated mining that has been environmentally destructive throughout the entire watershed, with incalculable consequences for riverine life.

In the last decade, the large-scale mining of sand and gravel has represented an unprecedented intervention into the watershed's landscape. Such bulk mining is inconceivable apart from advanced industrialization. The sand is dredged by heavy machinery, sorted into different grades for purposes as varied as the manufacture of concrete or glass fabrication, and transported on convoys of barges to be sold domestically or in neighboring countries as construction fill or road-building material as well as to provide well-drained foundations for new buildings. The micro and macro effects on the river's hydrology, bed, and the organisms that normally dwell there and in the riverine habitat as a whole are enormous and only recently fully appreciated. Sediment transport—that is, the sediment pulse—is a natural function of the river, but in this case, sediment transport becomes a capitalist commodity marketed locally and abroad. One might think of it as the ultimate step in disassembling and slicing up a natural assemblage and process that is "a river" into discrete natural resources that can be harvested and sold. Fish, arthropods (shrimp), and molluscs have long been captured and sold; the water itself is harvested and sold for irrigation; and now even the bed and banks of the river are scooped up and sold as commodities.

Sand mining.

Over the past few decades, the water flowing in the Ayeyarwady and its tributaries has become far more toxic and polluted than previously. Burma's comparatively low level of industrialization has, until recently, postponed some forms of industrial contamination that have plagued more developed watersheds such as the Mekong.

The term *industrial pollution* applies quite as much to modern agriculture as it does to industry per se. Many of the agricultural soils of Burma, though fertile in many respects, are deficient in nitrogen and phosphorus, both key ingredients in industrially produced fertilizer, now imported in large quantities. Continuous, nonfallow cultivation of the same plot with padi, pulses, or even legumes requires repeated applications of fertilizer if yields are to be maintained. The result, proportional to the expansion of riverine cropland, is a basinwide runoff of fertilizer residue that, when added to untreated sewage and massive amounts of organic waste emitted by factories, creates seasonal dead zones. Such events are common whenever a river is

Insecticides	Human health	Aquatic organisms	Aquatic organisms	Aquatic organisms	Terrestrial organisms	Terrestrial organisms	Terrestrial organisms
	Mammals	Fish	Crustaceans	Algae	Birds	Honey bees	Earthworms
Chlorpyrifos	Very High	Very High	High	Moderate	High	High	Moderate
Acephate	Moderate	Moderate	Moderate	Low	Low	Moderate	Low
Cypermethrin	Moderate	High	High	Moderate	Low	High	Moderate
Carbofuran	High	Moderate	High	Moderate	High	High	Moderate
Lambda-Cyhalothrin	High	High	Moderate	Moderate	Low	High	Moderate
Dimethoate	Moderate	Moderate	Moderate	Low	High	High	Moderate
Chlorpyrifos + Cypermethrin	Very High	Very High	High	High	High	High	High
Imidacloprid	Moderate	Moderate	Moderate	Low	High	High	Moderate

Insecticide risk level based on use and toxicity.

used as an outlet for untreated industrial wastes. The volume of organic wastes simply overwhelms the oxidizing capacity of the river water, thereby endangering all those life-forms that depend on a certain level of oxygenated water—many fish, crustaceans, insect larvae, and nymphs.

The industrial Ayeyarwady is increasingly poisoned more directly by toxic substances above and beyond the hypoxia caused by organic matter. Salts, metals, disinfectants, glues, resins, dyes, polychlorinated biphenyls (PCBs for electric circuits, cables, and lubricants), and formaldehyde are among the most prominent toxic pollutants that increasingly contaminate the Ayeyarwady. Most of these toxic substances are, of course, found downstream from centers of manufacturing in Mandalay, Sagaing, Pokokku, Monywa, and throughout the Yangon metropolitan area. Less concentrated, but perhaps even more deadly to river life along the watershed, are the waste products of chemically assisted industrial agriculture. Since the opening of the economy after 2010, there has been a huge increase in the importation and application of pesticides. Rice yields are threatened by nematodes (an invertebrate roundworm among the most numerous multicellular organisms on the planet), and the use of organophosphate insecticides has become general wherever padi is grown. Herbicides, fungicides, fumigants, repellants, and rodenticides have also come into general use for other crops grown in the watershed. It is, of course, a rare pesticide that *only* damages the immediate target at which it is directed and not other life-forms with which it comes into contact. In this case, great quantities of pesticide residues are washed into the river to threaten much of the aquatic life, mammals, and birds that depend on uncontaminated flora and fauna for their nutrition. Thanks to a comprehensive study of the state of the Ayeyarwady Basin, we have an estimate of the relative risk entailed by the main insecticides in wide use.

FIVE

Nonhuman Species

> Water is the elemental riverine "species." Does it not therefore
> have inalienable visitation rights to the wetlands, swamps,
> floodplain that are its offspring?
> —JAMES C. SCOTT

Ayeyarwady river dolphin.

AYEYARWADY RIVER DOLPHIN
(*Orcaella brevirostris*)

I belong to a gravely endangered species, as you may know. I was asked by my riverine friends (well, most of them are friends) to speak for them and to intrude on your single-species town meeting. I apologize for the disruption, but I and those I represent form most of the riverine life, especially those whose lifeworld is *in the river itself* and its floodplain. Why were we not invited in the first place? For us to be excluded from a discussion of the river is both arrogant and bound to result in a woefully incomplete understanding of the life-giving properties of its waters. We aim, by bursting in in this fashion, to broaden and, indeed, democratize the conversation.

Why was I chosen to speak on their behalf? Well, that should be obvious. I'm a mammal, and a charismatic one at that, who lives *in* the river. Whenever the discussion among humans turns to the endangered species in the river basin, we dolphins are among the species first mentioned. If I may speak frankly, my riverine companions and I also believed that you, being mammals yourselves, would be more likely to lend an ear to another mammal like me than to, say, birds, fish, reptiles, amphibians, crustaceans, molluscs, or insects, let alone plants, algae, or microbes, who, together, represent the vast majority of our river citizens. I speak, therefore, more on their behalf than for my own species.

What my riverine companions and I have to say is, of course, poached from the records of generations of naturalists, botanists, zoologists, ethnologists, biologists, fishers, hunters, gatherers, farmers, and others who have tried to make sense of our lives, habits, histories, and connections. So it is, in the end, the findings of *Homo sapiens* (the curious, the scientists, the engineers, and both the protectors and the predators) who are putting words in my mouth, not to mention the author who has assembled the words on this page. This is fitting because you humans are also riverine citizens. You are therefore entitled to a voice. But in recent decades, you have behaved as if you were the sole owners of the river basin and thus entitled, with the powerful tools at your disposal, to shape it for *your* short-term benefit.

We have devised a plan for presenting the view of nonhuman riverine life. Let's imagine that, at each and every meeting of the communities along the river convened by the geologist authors of an otherwise scientifically detailed report on geology and hydrology, nonhumans were present to give voice to their experience and interests—that is, to talk back.[1] We aim to appeal to our human counterparts by emphasizing species they know best and that—in the case of fish especially—form a major part of their subsistence. But we insist as well on representing lesser-known species, not only molluscs, birds, amphibians, and reptiles but also the multitudinous species of insects, microbes, and phytoplankton that form the very basis for sustaining the pyramid that is the living river. These last never appear in any cost-benefit analysis as they are not market goods, but without them, the river would die.

I apologize for the inconvenience of flooding your community hall with water that comes up to your waist. The river *nats*, whom we also revere, arranged this temporary flooding so that our river citizens could participate on equal terms—air for you, water for us. Fear not! Once our meeting is concluded, the *nats* will use their powers to spirit the water away.

The authors of the massive study of the Ayeyarwady's geology and hydrology organized community meetings, starting at the headwaters and proceeding zone by zone downstream, soliciting concerns about the river from riverine towns and villages. They then summarized (statistically) those concerns, often illustrating them with direct quotes. We, the more-than-human riverine citizens, will take those quotes as a point of departure and ask one affected species to offer its perspective on these human concerns. Each of our innumerable nonhuman species experiences a different river, and thus our opinions are often as varied and contradictory as are those of humans about the river.

Snow carp.

Snow Carp
(Tor yingjiangensis, Also Known as Tor Barb)

Dump soil from gold and jade mining should be carefully thrown, supervised by the government.
—FARMER FROM SHITEPU NEAR MYITKYINA

We can't survive long if mining continues at the current pace. We were singled out by you humans as a distinct species only twenty years ago, and you know relatively little about us except that we are threatened by pollution. Our numbers are plummeting, devastated by the cyanide and mercury used to separate the gold from the slurry—and, above all, by the vast amount of sediment turned loose by erosion and placer mining (gasoline-powered hydraulic hoses are used to dislodge soil) that radically changes the water quality, the bed of the river, and, over time, its course. We require clear, oxygenated, and nutritious cold water, and mining has raised the water temperature and so muddied the water, making feeding increasingly difficult. Gone are many of our favorite spawning sites, buried in sand and silt. Gone are many of the plants and insects that were the staples of our diet. I know from my migrating piscine friends that mining like this is widespread along both the Ayeyarwady and many of its smaller tributaries. Mining at the current scale, even if more closely regulated, will doom us as a species.

I understand that you humans (or at least many of you) value clean water for drinking and a healthy population of fish that will reproduce as fast as you can eat them. But your response to the declining fish catch has been devastating for us, *Tor yingjiangensis,* and in the long run, it will be devastating for you as well. Many of you have resorted to electrofishing, using battery-powered shocks to stun us, or to using cyanide to incapacitate us. Both of these methods indiscriminately kill other species as well (insects, plants, amphibians, reptiles, and so on). And even those of you who reject such methods are still likely to use nets of finer and finer mesh that trap fingerlings before they can spawn. These techniques are now practiced throughout the basin. They threaten our very existence; you may temporarily stave off the decline of the fish catch but you are, in effect, eating your own (riverine) seed corn.

Hilsa.

Hilsa

(*Tenualosa ilisha*)

> If we catch the small fish, of course then they can't grow bigger.
> But if I don't catch them, I can't make a living.
> —VILLAGE FISHER IN KYAUK TIN

I appreciate your anxieties about making a living, but it is not clear that we, the hilsa community, will live at all. We understand that we are not endemic to Burma alone (we are a staple of the diet in neighboring Bangladesh), but we are nevertheless the most desired fish species in the Burmese diet. As a species of the shad family, we are packed with nutritious omega 3 fatty acids. But there are fewer and fewer of us and most of us never manage to mature and spawn. Soon we may be gone altogether.

Yes, you are responsible for using nets of finer and finer mesh and some of us are electrocuted and poisoned in your passion to obtain a large catch to sell to the fish traders. But even without your predations we would be in deep trouble. You see, we move sideways, following the flood pulse onto the floodplain. We and many other species are migratory by nature; we can't reproduce if we can't move both vertically and laterally with the flood pulse. We move in harmony with the movement of the river itself and with the seasons. Some of us move hundreds of kilometers upstream at the beginning of the wet season and most of us move sideways as the flood pulse permits, so we can spawn on the nutrient-rich floodplain and the ephemeral wetlands and ponds that the annual floodwaters refresh. Our progeny then slowly grow to maturity in the relatively safe backwaters until they are big enough to return to the river and its major tributaries and, eventually, out to sea. In that way we fulfill our life cycle and replenish our kind.

Or . . . that is what we have done for millennia and what we desire to do today. But in countless ways over the past several decades, you have penned us in; you have incarcerated us and prevented us from reproducing. That's the main reason why you are chasing fewer and fewer of us. We depend *absolutely* on the in-between riverine landscape—sometimes wet and sometimes dry—from which we are now largely barred. In the upper reaches of the Ayeyarwady's tributaries, your government has built dams that block our migration routes and scramble all the signals of flow and temperature that we have relied on to begin and end our migrations. The many dams on the tributaries have totally disrupted our life cycle and that of others, especially the snow trout and eels. Although we are historically a species that moves between saltwater and freshwater spawning, these dams have trapped many of us.

So, my *Homo sapiens* friends, even if you stopped overfishing and using electric shock and poisons, we would still be in danger of extinction. The biggest threat to us is your dedication to agriculture, especially rice, for which you build dikes and levees to protect your fields from floods, denying us the wetlands we need for spawning. You then build sluices to direct water to your cropland, around which we cannot navigate. You have, we are told, drained three-quarters of the floodplain wetland in the entire basin, initiating a Great Drying, forcibly separating the river water from its natural floodplain. And then you are surprised and disappointed when there are fewer and fewer of us. Though we would be content merely to be left alone, I can't resist pointing out that, once again, you are eating your own seed corn, threatening the fish so essential to your diet. We, thanks to what you consider the delicious hilsa curry, provide most of the essential protein, minerals, and vitamins in your diet. Without us, the shrimp, and the fish paste that is your most important condiment, your carbohydrate-heavy diet would be unsustainable. Without us you would perish; you can flourish only if we flourish too.

Ayeyarwady River Dolphin

We depend on rainwater harvesting because the river water is dirty.
—Villager in the Lower Ayeyarwady

Fifteen years ago, the water quality used to be very good, compared to last year's.
—Assistant Village Health Officer

That's my cue to redirect the conversation. Speaking as an aquatic mammal and as the appointed representative of all forms of riverine life, I must remind you that the vast majority of riverine citizens are those you seldom, if ever, acknowledge, let alone understand. I speak, among others, of the riverine plants, insects, algae, molluscs, worms, larvae, zooplankton, and countless millions of microbes. When you *Homo sapiens* talk about the river, you understandably talk about water quality, flooding, and irrigation for your crops. When you do talk about river creatures, you dwell on fish, either those you want to catch and sell or those you like to taste in your curry or *mohinga*. I confess that as a large, charismatic aquatic mammal and one that actually often herds fish into your waiting nets, we dolphins have, by virtue of being on the cusp of extinction, benefited from the attention of the worldwide conservation movement.

Fish, and above all commercially valuable fish, tend to dominate conversations about the state of our river when it is not exclusively about even narrower human concerns such as the danger of floods and contaminated drinking water.

You eat fish. What do fish eat? The answer depends on the fish; some eat only plant matter; some eat other fish and zooplankton;

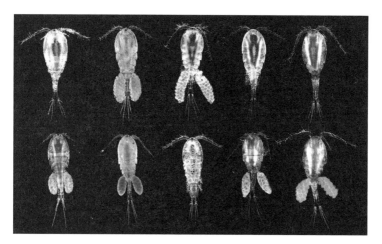

Copepods.

and most are omnivores, eating both. Some are bottom feeders, combing the riverbed for nutrients (that is, worms and minute crustaceans such as copepods).

The health of the river, the quality of the water, the health of the fish you eat, and ultimately your health as well depend on the health of this massive swarm of nearly invisible creatures that form the basis of the river's food chain pyramid. Let me introduce you to a couple of creatures near the base of this pyramid that nearly always fall outside your terrestrial gaze.

White ginger.

White Ginger
(Curcuma candida)

Like many other plants, I am quite hardy. All I need to "eat" is sunlight and, as long as my root system is largely intact, I can recover from most injuries. But, like other plants as well, I cannot flee danger by swimming or flying away as fish and birds can. To move, to colonize new areas, I depend on the movement of the current, herbivorous fish, and crustaceans to carry my seeds.

Unlike more mobile creatures, we aquatic plants are dependent in the short run on the conditions where we have taken root. In particular, we are creatures of the in-between. To thrive, we need quasi-dry conditions for photosynthesis and the rich, nutritious mud brought to us annually by the flood pulse. Swamps and marshes are our preferred habitat. Come to think of it, our lives are comparable to yours, humans: our flowers ("heads") are above the water and our roots ("feet") are underwater. We are the personification of in-between organisms. The combination of deforestation and drainage designed to turn ephemeral wetlands into cropland and the erection of dikes to block floodwaters has eliminated most of our habitat. It was not just our habitat, but one shared with spawning fish and waterfowl; it was a safe refuge for newly hatched fish to grow. Though we were an important food source for many species, we nonetheless prospered together. The Ayeyarwady version of the Great Drying over the past century has virtually finished us off.

More recently, our remaining numbers have been assaulted by industrial-scale incursions into our habitat. We are thus judged to be vulnerable to extinction in the near future. Large-scale sand and gravel mining has buried us under sediment, changed the composition and form of the riverbed, and polluted most of what remained of our habitat. The threats we face are comparable to those of other creatures that cannot run away from danger and seek a more secure habitat elsewhere. Let me, then, introduce a fellow species also handicapped by relative immobility—and another whose condition and fate are ignored by *Homo sapiens*.

Mollusc.

MOLLUSC
(*Lamellidens mainwaringi*)

Much of our fate is shared with that of white ginger, although we rely less on the in-between and more on the riverbed. We are a numerous species. Many of us are endemic along the Ayeyarwady and its tributaries, and we are now threatened especially by what humankind has done to our habitat. The sediment and poisonous chemicals associated with sand, gold, rare earth metals, and jade and titanium mining have either buried or poisoned many of us. We are also threatened with extinction by modern agricultural practices—the runoff of fertilizer waste, insecticides, herbicides, fungicides—not to mention the chemicals used in manufacturing and then dumped into the river along with plastics and sewage. We, along with wetland plants such as white ginger, have, by our filtering and fixation of pollutants, helped keep the water relatively clean. Now that our very existence is at stake, we can no longer clean up your mess.

Ayeyarwady River Dolphin

Thus far, the riverine citizens who have intruded on your single-species conversation to make their voices heard are creatures who live *in* the water. That's why you have been up to your waists in (polluted) water. A few of us water dwellers can breathe—such as dolphins, who have lungs, or those who engage in gas exchange by breathing through their skin and mouth lining, such as mudskippers. We are not, however, the only riverine creatures. Many others, not unlike many of you humans, depend vitally on the river for their subsistence but do not dwell in it. As an example, let me introduce you to the oriental darter.[2]

The darter, like the cormorant, has accomplished the amazing feat of mastering all four elements of the riverine environment: it can fly, it can walk on land, it can swim on the surface of the water, and it can dive beneath the surface for long periods in pursuit of the fish that form the bulk of its diet.

Oriental darter.

Oriental Darter
(Anhinga melanogaster)

We waterbirds and freshwater fish are, in many respects, in the same boat (no pun intended). We are among the most depleted and endangered worldwide. Like many fish, we depend on the flood pulse to refresh the ephemeral wetlands that are our most important hunting grounds. We, like other riverine species, including yourselves, depend not only on our immediate habitat but on the habitat of our prey, the source of our nutrition. The drainage and cultivation of this habitat have made it difficult for us to thrive and reproduce. That's only the beginning of our troubles. We depend on nesting places in shaded groves along the banks and nursery backwaters for our fledglings; many of these have been destroyed by deforestation and widespread erosion. Massive sand mining in the last decade has only made our situation more dire.

Once upon a time our biggest worry was being hunted for our meat and plumage, but that threat has been overwhelmed by industrial effects. Perhaps the most dangerous of these human effects are the poisons used in gold mining and industrial agriculture. Cyanide, used to fix gold throughout the watershed, has damaged our capacity to reproduce. There are agricultural chemicals, including insecticides such as chlorpyrifos, that threaten us as well as fish, insects, shrimp, and other zooplankton. Banned in much of the world, chlorpyrifos is extremely toxic to all forms of riverine life. In a larger sense, the health and survival of all riverine creatures depend on uncontaminated, clean water. Most of the increasingly lethal contamination of the watershed over the past several decades is due exclusively to human activity: industrial and plantation agriculture (herbicides, insecticides, fungicides), mining, manufacturing waste, and human sewage. We, unlike molluscs, have the gift of flight and can therefore move to another watershed, but we worry that, before you humans come to appreciate that you are fouling your own nest, we will have all died or fled the Ayeyarwady watershed.

Ayeyarwady River Dolphin

Let me introduce you to the Burmese roofed turtle (*Batagur trivittata*), a water-dwelling reptile once thought to be extinct but rediscovered in small numbers, although still critically endangered.

Burmese roofed or smiling turtle.

BURMESE ROOFED TURTLE OR SMILING TURTLE
(*Batagur trivittata*)

There are only few of us, and we are endemic to this watershed alone. We are reptiles, though, like frogs, we dwell in and by watercourses. Like frogs, we are creatures of the in-between, the periodic wetlands refreshed annually by the monsoon rains and floods. The overriding reason why we are in grave danger is simply the loss of our ecological niche. Humankind has occupied our habitat, felled the trees, drained the water, and diked and planted the places where we thrived. Our eggs were hunted avidly by birds, mammals, and humans along an increasingly denuded and populated riverbank. Our meat, shells, and vital organs were prized both for consumption and their supposed medicinal properties. One of the first signs that we were not, in fact, extinct was that some of us were discovered for sale in China in 2002! Prior to 1940 it seems that we were abundant in the wild, but now, owing to the devastation of our habitat and poaching, most of us who survive are protected, as I am, in conservation sites like the Yadanabon Zoological Garden in Mandalay. Yet we are still hunted for our shells, for our supposed medicinal properties, and as pets.

We suffer as well from the growing pollution of the Ayeyarwady—not only the poisons but also the plastic waste we inadvertently ingest along with our diet of plants, worms, snails, insects, and fish. We, along with our amphibian cousins the frogs and salamanders, are endangered not only in the Ayeyarwady Basin but worldwide because so much of our freshwater wetland habitat has been turned into dry land.

Ayeyarwady River Dolphin

So far, except for myself, of course, you have only heard the plaints of nonmammals. We worry that you are socialized to pay far less attention to the losses of those species who do not share your traits of warm blood, air breathing, and fur. So, as a final witness, I introduce a pair of closely related mammals: the Asian hairy-nosed otter and the Asian small-clawed otter, who, like the dolphin, are charismatic and playful.

Asian Hairy-Nosed Otter
(*Lutra sumatrana*) and the Asian Small-Clawed Otter
(*Aonyx cinereus*)

I am a hairy-nosed otter. I'm speaking on my behalf and on behalf of the smallest otter: the small-clawed otter. It is claimed that we are probably extinct in Burma, but this conclusion is because in all of Southeast Asia, Burma is the least investigated and surveyed for rare species. Burma is, after all, an isolated and

Asian small-clawed otter.

war-torn country, and we otters prefer lowland forested swamps, coastal mangroves, and tidal pools and wetland habitats that are often difficult to penetrate. We are almost certainly present in small numbers; we're simply unobserved. Why the statistics for endangerment of species are calculated by arbitrary national boundaries rather than habitat is beyond me.

It's true that we are charismatic in the opinion of *Homo sapiens,* but that has been more of a curse than a blessing. We, like pangolins, have been valued in the sophisticated, illegal, worldwide network of wildlife trafficking. In particular, we are valued for our fur, our meat, and increasingly as pets. It is also our misfortune that the civil war that has erupted in Burma since the military coup of February 2021 has greatly increased the volume of poaching. Human discord is of no concern to otters, but that does not spare us from the depredation that is occurring within the confines of those arbitrary boundaries. Hard-pressed rebels and villagers, lacking more lucrative and legitimate pursuits, have little choice but to traffic in whatever prey they can hunt, knowing that there is a well-established network of middlemen and itinerant merchants that can move us, dead or alive, to the most lucrative markets in China, Vietnam, and Japan.

The greatest danger we face, however, is the one mentioned by virtually everyone here: the loss of our habitat. We thrive best in the slow-moving, meandering river, in ponds, swamps, mangrove coasts, and seasonal wetlands. So much of this habitat has been deforested, drained, diked off from the river, replaced by aquaculture, and occupied by the infrastructure of crop-planting humans that we are confined to what little patches of our habitat remain. Our prey—small fish, mudskippers, crustaceans, molluscs, and snakes—are harder to find, and those that have survived have become increasingly toxic thanks to agricultural pesticides. Though we are, as mammals, more mobile than fish, it is far more difficult for us to seek out a new watershed; we envy birds and their power of flight.

Ayeyarwady River Dolphin

We nonhumans who have spoken demand our full rights as citizens of the watershed. We recognize that you too are riverine citizens; you desire clean water and a reliable and abundant harvest of fish as part of your subsistence. But both are in jeopardy because of what you and others of your species have been doing. What we have here is a world-historic land and water grab in which a single species has seized an entire landscape from its indigenous inhabitants and unilaterally *colonized* it. We are abject colonial subjects just like indigenous peoples whose land was appropriated by imperial expansion. We are, in much the same way, the subjects of a quasi-universal settler colonialism. As comparative newcomers in evolutionary terms, you declared the territory aqua nullius and terra nullius—water and land belonging to no one and therefore open to your claim of exclusive sovereignty over all forms of life.

You not only defied any claim we might have had to our habitat, you took the apocalyptic step of obliterating the landscape itself. Where there had once been fluvial forests, you deforested, drained, and planted the land, destroying wetlands, swamps, and the in-between, ushering in the Great Drying. Where there were floodplains, you built levees and weirs to keep the river in its narrowest channel. Where there was free-flowing water, you erected hydroelectric and irrigation dams that took the essential *riverness* out of the river and substituted a chain of lakes. Where the river carved out a shifting habitat depending on the monsoon, your mining, dredging, and infrastructure totally changed the hydrology of the water flow and the movement of sediment. And where the water was relatively clean, you used it as a toxic sewage pipe filled with plastics, human excrement, and agricultural and industrial waste.

We want our river back from the colonists! We want its floods, silt, wetlands, swamps, mangrove forests: in short, we demand an all-species riverine democracy as the essential condition of our existence.

SIX

Iatrogenic Effects

Hunter-gatherers pluck what they desire from a teeming, biodiverse world. Over time, we know that premodern hunter-gatherers tend to modify their immediate environment to enhance the availability of fruits and edible and medicinal plants and to attract game. Beyond hoes and hand labor, their major tool in this endeavor is fire. This long, drawn-out process has been termed slow horticulture. It features changes and adaptations, but its aggregate effects are marginal. Premodern hunter-gatherers might slightly reduce the radius of a meal, but they remain creatures of movement whose subsistence depends on following the rhythms of the undomesticated world.

The modern agriculturalist or pastoralist, on the other hand, plucks from the natural world the most desirable and amenable cultivars and mammals and then segregates them from the natural world in order to shape them to best serve the needs of *Homo sapiens*. One result of this process is a kind of horticultural and mammalian apartheid, fostering, over time, subspecies of plants and animals that are co-evolved with humankind such that they can no longer survive on their own and must rely on human protection. This is the essence of what we understand as "domestication."

The early state could not exist without a concentrated sedentary farming population growing domesticated cereal crops. Domesticated cereals were not an invention of the state. The long process of domestication was completed several millennia before anything like a state was created around 6500 BCE. Putting it epigraphically, there is sedentary grain growing *without* the state but there is no early state without sedentary farming.[1] Sedentary cereal agriculture is better seen as the necessary *scaffolding* for the formation of the early state.

Prevailing forms of subsistence that predate fixed-field cereal cultivation (and still persist), such as hunting and gathering, shifting agriculture (swidden or fire-field agriculture), pastoralism, mixed marshland subsistence, and the cultivation of roots and tubers militate against concentration, against appropriation from above or outside, and against assessment and surveillance of production and surpluses. The relation between state formation and sedentary cereal agriculture conforms to what Max Weber termed elective affinity.

The early state is an ecologically invasive, artificial order. Fixed-field agriculture, irrigated wet rice, and aboveground simultaneously ripening cereals all require a simplification of the landscape, while the concentration of cultivators and storable subsistence goods are ideal for the process of early centralized state-making. The state is constantly modifying its environment, simplifying it within the means at its disposal and within the limits of what nature will absorb without fighting back—or, more frequently, dying.

The central paradox behind this concentration and simplification, as we shall see, is a domesticated landscape that is politically, ecologically, and epidemiologically extremely fragile.[2]

Whereas the hunter-gatherer *adapts* to the complex rhythms of the natural world to subsist, the early state strives to subdue this movement and complexity—to create a *state-serving habitat.* A state-serving habitat is one stripped down to a narrow band of domesticated crops and domesticated animals concentrated within a small radius to facilitate appropriation. This implies an aspiration (never completely realized) to *stasis,* to solidifying that map-based illusion of a sharp border between the river and the floodplain. It means attempting to abolish the in-between; turning mud, as it were, into land or water. This process is captured by the iconic Dutch windmill, whose function is to drain marshland and transform it into cropland or pasture.[3] A group of investors would typically finance the windmill on the understanding that the hypothetical cropland would be apportioned on the basis of how much each investor contributed; the result was landed property for the investors, the concentration of production and population, and revenue for the state. Above all, in our terms, this process necessarily entails, wherever possible, the replacement of wetlands, old-growth forests, and complex, patchy, biodiverse ecosystems with one or a few annual crops and a few species of domesticated livestock. Precisely because these domesticates cannot survive in the wild, they must be protected—cordoned off from nature—from wild nature, which is *stigmatized as weeds, wastelands, varmints, pests,* or *predators.* A weed, of course, is simply a plant we judge to be out of place and that competes with our domesticated cultivars. Varmints (such as rats or wolves) are mammals that feast on our grain or on our domesticated livestock. And, to top it off, the term for other *Homo sapiens* who are not cultivating state subjects and who, in fact, may raid agricultural settlements, is *barbarians*.

The term *domesticated* in its strict sense as control over reproduction applies most directly to cultivars and domestic livestock. It applies, I believe, in the early state to slavery and patriarchy as well, where reproduction is regulated and women and slaves are property, much as cattle are. When it comes to living beings—even domesticates—total domination is aspirational; it is never fully realized. For a complex, ever-moving entity like a river, the most that can be achieved is better represented by terms like *regulation, control, discipline,* and *pacification*. Maxim Gorky, a fan of Stalinist utopian visions of total control, described the process as one of making "mad rivers sane." In this case, the aim is to harness rivers to do our bidding and serve our ends.

In what follows, we shall first trace the process of taming nature in agriculture, forestry, industrial crops, and human labor itself. Then we shall examine the process in the domestication of livestock and aquaculture. Finally, we will examine the paradoxical and frustrating efforts of states to discipline rivers to serve our narrow purposes.

Agriculture and Forestry

As Timothy Weiskel has observed, agriculture depends on ecocide.

Nothing has so transformed the world's landscape as the practice of agriculture. The process would have been visible in select places by the second millennium BCE, but the results were small and inconsequential—a mere rounding error in terms of area.

And yet, all the essential elements of the process were in place: the selection of one or another cultivar as a staple in the diet, the use of fire to clear land for crops and pasture and, haltingly, the growth of the population living essentially by agriculture. It became the hegemonic mode of subsistence throughout the world until the industrial revolution, when fossil fuel power, artificial fertilizers, and exponential population growth powered it to unforeseen levels.

According to one rough estimate, something like two hundred thousand plant species have been identified.[4] What is astounding is that from this vast floral cornucopia only a dozen (0.00006 of the total) account for 80 percent of the world's food consumption. The lucky (?) domesticates, with humans on hand to tend to them (cultivating, weeding, watering) and to fend off birds, livestock, and wild predators, are maize, sorghum, rice, wheat, barley, manioc, sweet potatoes, potatoes, sugarcane, sugar beets, bananas, and pulse sorghum.

Why do these plants so dominate agriculture and the world's diet? The cereals in particular—maize, barley, wheat, and sorghum—had properties that lent themselves to domestication. As largely members of the grass family, they were relatively hardy and their growth rate generally outpaced the weeds competing with them. This last feature was important because the earliest forms of agriculture were not plow agriculture but rather flood-recession agriculture and shifting agriculture. In the former, the field is cleared by floodwaters that, as they recede, leave behind a rich layer of silt, whereas in the latter, the field is cleared by fire, leaving behind a layer of nutritious ash. If the cultivars are immediately sown, they stay ahead of (for a time) the weeds that compete with them for nutrients, water, and sunlight.

A second vital characteristic of the cereals was their capacity, evolved over time, to resist shedding their seeds until they arrived at the threshing floor. In this context, it is worth remembering that all the precursors of crops were first gathered where they grew in wild stands. Normally, wild plants shed their seeds (dehisce) sporadically. It is an effective strategy for plant propagation in the wild but limits the proportion of the cereal that makes it to the threshing floor. As near as we can tell, the process of *indehiscence* was a long-run result of gathering desirable wild cereals, an example of quasi-unintentional domestication. Over long periods of time the sheaves of grain that did make it back to the domus held on to their seeds simply because they did not shatter in the field while being gathered. Once the slow process of domestication was underway, the early planters had a cultivar that promised a substantial yield. Much the same process of indehiscence applied to the other cereals.

The cereals in particular had other characteristics that made them valuable domesticates: they are (with the exception of maize) self-pollinating (that is, breed true); they are fast growing and hardy; they can be dried; and they store well and can be transported relatively long distances with higher value per unit of weight and volume than other cultivars (for example, potatoes and manioc). These last two characteristics are decisive; these are the crops that sustained early states.[5]

Cereals are dense in terms of caloric yield per unit of land and also in terms of the demographic concentration of cultivators that early state-making required. But there are other early domesticates that have high food value and can be dried and stored successfully. Why are there no legume states in the historical record relying upon, say, lentils, chickpeas, or peas?[6] Why are there no taro or soybean states, no sago or breadfruit states, no yam, cassava, banana, peanut, or potato states? Singly or in combination, they would provide for the subsistence needs of a sizeable population. The key to the nexus between cereals and states lies, I believe, in the fact that only cereal grains can serve as a basis for taxation and appropriation: they grow aboveground; their yield can be assessed; they are storable, portable, and can be used as rations; *and most of the crop ripens at the same time.* This last characteristic is crucial, for it means that tax and tithe gatherers can appropriate all or part of the harvest at a moment's notice, either in the field or from the granary after it has been threshed.

Reliance on an extremely small bandwidth of plants carries with it tremendous dangers. The crowding of any genetically similar species—be it *Homo sapiens,* a staple crop, or domesticated animals—turns out to be a feast for birds, insects, bacteria, viruses, fungi, or rusts that are adapted to that particular species. This combination of genetic similarity, crowding, and epidemiological vulnerability helps explain why the early states were plagued with new infectious diseases affecting the human population, their livestock, and their crops. The narrower the genetic variety in the major staple crop, the greater the state's vulnerability. In addition, in the Mesopotamian basin, wheat was gradually displaced by barley, as the repeated cultivation of wheat contributed to the salinity of the soil, which barley tolerates better than wheat.

There is no way the earliest cultivators of fixed-field crops could have realized that repeated annual planting of the same crop was depleting the soil of vital nutrients, except through the experience of declining yields. Over time, techniques like fallowing mitigated the loss but required more land.[7] The most striking example of soil depletion, of course, is plantation agriculture, in which the cultivars are often genetically identical clones and thus equally vulnerable to the same pathogens.

For our purposes, however, the overriding impact of the spread of agriculture was the wholesale remaking of the landscape. It required deforestation and drainage. At the very earliest stages, when production was low and clustered near rivers, the effect on riverine mammals, birds, amphibians, insects, and crustaceans was apparent but modest. Forests were seen as sources of firewood, charcoal, and construction material—that is, tools at the disposal of *Homo sapiens,* not the essential habitat of birds and insects. Swamps and wetlands, the most biodiverse habitats, were seen largely as wastelands—land that in its present state served no human purpose—and, for that matter, were seen as dark and threatening places.[8] Draining swamps became not only a political slogan but a source of civilizational transformation. It virtually defined Dutch identity until recently. The draining of the English Fens, Mussolini's draining of the Pontine Marshes, and Saddam Hussein's draining of the Mesopotamian Marshes are only three examples of what has been a worldwide process. Environmentally, it should be clear by now what was happening. A complex habitat suited to millions of birds, insects, and fish—among others—has been drained, simplified, dried, and plowed to favor annual cereals and thus destroyed, largely, as a habitat for most other species. In the case of wetlands worldwide, this massive project of environmental engineering has resulted in the loss of fully half of the globe's terrestrial wetlands. The Swampland Act of 1850 in the United States transferred swamplands to state ownership with the proviso that all proceeds from the sale of such lands would be used to promote agriculture. Although a few voices at the time cautioned about the hydrological consequences of such interventions, they were largely concerned not about the degradation of the natural world but about the human consequences in terms of floods and the drying of the land.[9] To be fair, our knowledge of the vast natural world and its connective tissue was such that, until recently, we had little idea about the devastation we were wreaking. For the past four decades, at least, however, that devastation has become a glaring reality.

The fragmentary areas where one can still find a rich, diverse collection of species are diagnostic. They are precisely those areas that have been relatively untouched by human landscape engineering: remote mountainous regions, national parks, bioreserves, and swamps and marshes that have not yet been drained. Mammals (such as bears and wolves) that once roamed the plains have retreated to safer refuges, in much the same way endangered humans have retreated to the hills and swamps to escape capture or death. If one is looking for insects and birds that have disappeared from populated areas, they too are more likely to be found in such out-of-the-way places.

The logic of landscape transformation is *utilitarian*—a natural resource (the shorthand is "land") is, as in neoclassical economics, something that can be combined with labor and capital to produce a commodity. Implicitly or explicitly, all other considerations about resources are stripped away, and the logic of maximum return for least cost prevails. The invention of scientific forestry in Germany in the late eighteenth century is a striking case in point.[10] The goal was to optimize forest-based revenue, as measured largely by the yield in cubic meters of timber and firewood. Hitherto, the revenue from domainal forests fluctuated widely depending upon the mix of tree species extracted and the terrain. It was the next step in simplification, however, that was decisive. The foresters set about careful seeding, planting, and cutting to create a forest that was easier to count, manipulate, measure, and assess in order to optimize revenue. They determined that the Norway spruce and the Scotch pine were, depending on the soil, the most profitable, fast-growing option. The new forests, on clear-cut land, were planted in straight rows like cultivated crops, and the result was a single-species forest, with trees all the same age for easy counting and felling. The new, rational scientific forest was at first very successful and became the world standard. Many attribute this early success to the accumulated nutrients still in the root channels of the old-growth forest that had been felled. In any event, after roughly a century, Germans had coined a new term—*Waldsterben* (forest death)—to describe their disease-stricken scientific forests.

This is a story of landscape engineering in the service of only two commodities: timber and firewood. It was, of course, a disaster for the local peasantry who gathered edible and medicinal plants, grazed their livestock, collected fodder and bedding, and hunted, trapped, and fished in the old-growth forest. This cornucopia was greatly diminished in a single-species forest. And, given the utter lack of tree species diversity, the ranges of birds, mammals, and insects that had proliferated in the old-growth forest were reduced to a miniscule proportion of their earlier diversity. The landscape was sculpted to accommodate a one-commodity machine and all other creatures were impoverished.

The logic of the one-commodity machine repeated in analogous cases. Jan Douwe van der Ploeg provides an instructive example from the Andes.[11] There are, he notes, indigenous potato-planting practices he terms as craft-based. Peruvian farmers have a large variety of potato subspecies at their disposal. As their fields are arrayed in steep terrain at different altitudes, different orientations to sunlight, and different soil conditions, they carefully choose those cultivars that, based on long experience, will do well in a particular niche.

Crop scientists, on the other hand, began by conceiving of and then breeding the genotype of the ideal potato—in terms of taste, cooking qualities, fast growth, and caloric and vitamin content—and then set about sculpting the landscape so as to accommodate that ideal potato. That meant, rather like scientific forestry, making a uniform landscape for a single cultivar, featuring a fertilizer formula, an irrigation schedule, a pesticide formula, and regimented weeding. The landscape was remade for the ideal potato, making it more uniform and less diverse, as opposed to the craft method, which took the environment as a given and chose the particular variety that would thrive in that setting.

Domesticated Livestock and Aquaculture

Vaclav Smil, an environmentalist and historian, attempted to quantify the ratio between the metric tonnage of domesticated flesh in the world (chickens, cattle, and pigs) and the metric tonnage of wild flesh (elephants, whales, deer, or antelope), not counting *Homo sapiens*.[12] The ratio was nine to one in favor of domesticates. If you want to know how much of the undomesticated world remains, this is the only statistic you need. The calculations are methodologically complex and subject to dispute (for example, insects are excluded), but if the figure is even roughly accurate, it is deeply sobering. If, in addition, we add to the calculations the nearly 8 billion *Homo sapiens* as domesticates, the ratio surges to eighteen to one. *Homo sapiens* has selected those rare species most amenable to domestication and confined and bred them to the point at which, like domesticated plants, they become "basket cases" that can no longer survive on their own in the wild—this, along with control over reproduction, represents the very definition of domestication. Domesticated sheep and goats, for example, became smaller, more placid, less aware of their surroundings (as captives, they have a human protector), less given to fright and flight, and less sexually dimorphic. By constant in-breeding of the most placid sheep (the most aggressive were probably eaten first!) the very nature of the species was changed.

As with plants, it is important to recognize how devastating the process of domestication was for biodiversity. Out of roughly six thousand mammalian species only four—pigs, sheep, goats, and cattle—became the ones enshrined in our diet.[13] Whole landscapes were devoted to their needs for pasture. The reduction in avian flesh, however, was dominated by a single species: *Gallus gallus,* ancestor and contemporary of the modern chicken, was by far the predominant species selected for domestication. There are three *Gallus gallus* for every human being on the planet, virtually all of them in industrial production.

The model of German scientific forestry, which ignored everything about the forest except for its efficient yield of timber and firewood, was applied to domestic animals. They became one-commodity machines, and everything else about them was ignored. Prior to industrial production, pigs were used, for example, to help clear land for agriculture, as were turkeys. Early settlers in New England would pen pigs and turkeys on a plot of uncleared land, knowing the pig's fondness for roots and the turkey's taste for foliage. In a couple of years, the field was ready for planting—with no additional labor required of the settler. Now pigs are raised almost exclusively for their flesh. Sheep, of course, were valued for their wool and milk, though meat breeds predominated. In a few cases, subspecies have been bred to maximize a single commodity: the layer hen bred for the most efficient and profitable egg production is the Italian Leghorn, whereas a variety of Cornish/Plymouth Rock is favored as a broiler for meat production. The cattle breed favored for milk production is the Holstein, and the subspecies most efficient for meat production is the Black Angus.

Once a breed has become, in industrial production, a one-commodity machine, no effort is spared to minimize the cost of its production. This is impeccable capitalist logic: the lower the cost, other things being equal, the greater the profit. This means, in the case of chickens bred for meat, feeding the selected breed growth hormones for the most rapid growth possible and a body confirmation that emphasizes the most widely desirable tastes—for example, large breasts for the U.S. market. Many thousands of hours have been devoted to sculpting this object, a chicken, into the most efficient commodity machine achievable. One striking indication of this quest is the lifespan of the industrial chicken. While its wild cousin has a life expectancy of nine to fifteen years, the industrial chicken is slaughtered and sent to market after only seven to nine weeks. It is surely a triumph of commoditization when the product's lifespan is roughly 2 percent that of its wild cousins. For the industrial pig versus its wild progenitor, the contrast is slightly less radical but nonetheless striking.[14] Estimates of a wild pig's life expectancy vary widely, though the average is roughly ten years, while the industrial pig is slaughtered at between four and seven months of age. A great deal of experimentation and scientific research goes into the process. The first criterion is the choice and breeding of the subspecies most susceptible to such manipulation. Growth hormones and carefully calibrated feed ensure early maturity and hence a faster delivery of the desired commodity. Confinement (incarceration?) is an integral element in the strategy. Caged chickens and penned-in pigs help ensure that valuable calories devoted to meat production are not wasted in movement.

The insidious capitalist logic of farming and domestication, as we have seen, is devoted to a tiny number of amenable cultivars, a miniscule selection of mammals, and a handful of birds, most notably the chicken. This process has brought about a very, very curated world of domesticates destined for the human diet. More recently that logic has been brought to bear on fish in the aptly named aquaculture. Fish farming has become responsible over time for one-third of Burma's fish production and is one of its fastest-growing sectors, given the decline in yields from traditional fishing. The logic should, by now, be clear. First, a very narrow band of species is chosen, based on their capacity to tolerate crowding, their potential rates of growth, and their desirability in the diet of consumers. In Burma, the species selected are tilapia, originally from Mozambique (*Oreochromis mossambicus*), and the giant gourami (*Osphronemus goramy*).

The logic behind each of these choices is comparable to the choice of other domesticates. Each species is adaptable, can withstand crowding, and matures quickly, minimizing feed costs. In the case of the Mozambique tilapia, it was later crossbred with other tilapia to achieve a fillet size that consumers coveted. The giant gourami has at least two characteristics, aside from adaptability and fast growth, that make it an ideal domesticate. It is tolerant of brackish water and, like a few other species, is capable of breathing moist air for extended periods. One infers that, as a Southeast Asian species, it was a denizen of the in-between and thus required this ability to survive through dryer periods.

As with other forms of human-induced domestication, the process involves at least two steps: the selection of the species most amenable to confinement and the sculpting by selection, crossbreeding, and eventually genetic modification to fine-tune their suitability as a profitable commodity machine.

We might well infer from the domestication of the natural world that humankind plays a godlike role with respect to domesticates. This is largely what the monotheistic religions, Judaism, Christianity, and Islam, teach. From another and equally plausible perspective, however, humans are the captives and even

Mozambique tilapia.

Giant gourami.

the slaves of their domesticates. The very nature of the domesticated life we have extracted from the wild means that it needs constant care and guardianship. This is all the more vital regarding the care of vulnerable clones. Having entombed, as it were, these domesticates in our domus, much of our life as a species is devoted to their protection—food, fertilizer, fencing, cultivation, weeding, pest control, and so on. Who serves whom becomes, at this point, a metaphysical issue of perspective.

The Domesticated River

Mark Cioc has observed in *The Rhine*, "River systems become highly engineered, optimized, hydraulic machines." Historically, prior to the Neolithic, the human population moved with the rhythm of the largely undisturbed river; they followed fish, bird, and mammalian migrations that were themselves triggered by the flood pulse and monsoon rains. The entire integrated clockwork of creaturely movement was set in motion by "reading" the river. Imperceptibly at first, humankind began efforts to move water where they wanted it and to block water from areas where they didn't. Hence the appearance of small levees, weirs, irrigation sluices, and small drainage ditches to facilitate cropping. At the very least this represented a small effort to regulate and tame the natural disturbance of the river—the engine of its biodiversity— and to appropriate its services. It's virtually impossible to stop the movement of a river, although a chain of hydroelectric reservoirs comes reasonably close. An enslaved river is, after all, put to work for the purposes of the master. Depending on the particular use to which it is put, the river might be termed *tamed, domesticated, regulated, enclosed, reshaped, disciplined, harnessed,* or even *tidied up.* One or a combination of such terms seems appropriate to describe the domestication of rivers.

The attempt to domesticate rivers is decidedly not a unique product of modern engineering. In the sixteenth century, long before the industrial revolution, Pan Jixun, whom Mark Elvin describes as one of the greatest of Chinese river-tamers, reshaped the Yellow River for the state to master sediment transport: "In round numbers, he built 1,200,000 million feet of earthen embankment, 30,000 feet of stone embankment, stopped 139 breaches, constructed stone spillways, dredged 1,500 feet of riverbed, planted 830,000 willow trees to stabilize the tops of dikes, drove a large but unrecorded number of tree trunks as pilings under the embankment, and spent 500,000 ounces of silver and nearly 127, 000 piculs of rice. He unified the course of the Yellow River and confined it within so-called thread dikes."[15]

The huge expenditure of material, labor, and finance was astounding and goes to show how massive were efforts to master the Yellow River in service of the Ming Dynasty.

NAVIGATION

Perhaps the oldest and most consistent efforts to tame rivers were initiated to make them navigable for trade, exchange, and territorial control. Because floating goods by boat entails enormous savings in time and labor compared to transport overland by draft animals and carts, virtually all early states, small or large, arose next to navigable water. Of course, the transport advantage is greatest when transporting goods downstream rather than upstream, but in either case that advantage is between six and twelve times greater than with shipping the same goods by cart.[16] Environmental evidence for this advantage is decisive and embedded in the landscape. Every settlement of any size is in constant need of timber for cooking, heating, construction, and boatbuilding, and the easily available timber around the settlement is soon exhausted. The next easiest option, as we saw in chapter 2, is simply to fell trees upstream of the settlement and float the logs down to the settlement when the water level is high. The pattern of riverine deforestation is, in this sense, diagnostic.

The way to think about a "river of navigation" is to understand that the aspiration of river sculpting for navigation—an aspiration that is never fully realized—is to have it approximate a straight two-lane asphalt road. This requires uniform dredging so that the depth of the channel can ideally accommodate two-way traffic. A pure navigational canal should be straight as a die and eliminate meanders, which not only lengthen the voyage but, by their very nature, deposit silt, sand, and clay, clogging navigation channels. Here the aim of efficient transport is in direct conflict with the contribution of meanders to wetlands and diverse habitats.

A striking example of the attempt to tame and control rivers concerns the Rhine River, as embodied in Johann Gottfried Tulla's Enlightenment-inspired international treaty, dubbed the Rectification of the Rhine—as if it were the duty of engineers to correct the imperfections of the river. Mark Cioc's brilliant account of the process is unsurpassed.[17] In many respects, it was very much an effort to apply the one-commodity machine of scientific forestry to a river. Many of the meanders were eliminated, snags and boulders impeding navigation were removed, most of the adjacent wetlands were cordoned off, and the overall length of the river was reduced by more than one hundred kilometers. As might have been foreseen, the removal of snags and meanders was a disaster for wildlife, and eventually flash floods became more common and devastating. Nonetheless, Tulla had achieved an eminently navigable river that became a major artery of commerce and trade.[18]

As Friedrich Schiller wrote in 1793, "Who of us would not prefer to dwell on the ingenious disorder of a natural river landscape than on the trivial regularity of a straightened water course?"[19] Schiller is making an aesthetic observation, but the ecology of a river "disordered" in this fashion is at least as consequential.

HYDROELECTRIC POWER

The number of tasks a disciplined, regulated river is made to serve is nearly endless. Without exception, they each require a reengineering, as we have just seen in the case of navigation. Its success and the realization that transport by water without transshipment to smaller crafts or carts was a decisive advantage contributed to the "canal mania investment bubble" of the late eighteenth century in England and Wales, the premise of which was to connect watersheds. The building of the Erie Canal in the early 1800s was similarly designed to join the Hudson River to the Great Lakes. In those terms it was spectacularly successful, although fifty years later it had to compete with a growing railroad network.

A good many purposes that have been assigned to the engineered river require many of the same techniques. The purpose of navigation mandated a straightened channel, uniform depth achieved by dredging, and confinement by dikes so as to maintain the required width and depth. Many of the other purposes—to prevent floods or to remove sewage, industrial waste, and pesticides safely out to sea—require different conformations. The engineering dilemma most of these tasks presents is that, once the river enters a low-gradient floodplain, it is bound to deposit more sediment, thus raising the riverbed, increasing the danger of flooding and new meanders. A standard solution to this problem worldwide has been to increase the volume and speed of the river's flow by erecting more levees and dikes, creating spillways from tributaries to scour the bed of the river periodically and maintain its depth. (Think of flushing a toilet!) These techniques and their successes and failures are amply documented in the literature, and for that reason I elect to move to the most audacious attempt to destroy the river as such and transform a watershed into a chain of lakes, each of which is dammed to generate electricity.[20]

The hydroelectric river is the ultimate one-commodity, optimized hydraulic machine. Because of the requirements of hydropower for a certain volume of water falling a certain distance to activate a turbine, the vast majority of such dams are located in the upper reaches of the watershed, where the terrain is steep.[21] In the major part of Southeast Asia, the terrain suitable for most dams is in the realm of ethnic minorities, who are culturally distinct from the lowland, rice-planting peoples, adding a dimension of ethnic conflict to the coercion of forced displacement and the knowledge that the power generated is unlikely to benefit those in the immediate vicinity of the dam. The growing number of dams built and planned on the Ayeyarwady are largely located on the upper Chindwin, where the Chin people live, and upstream from the Myitsone along the Maikha and Malikha, where the Kachin reside.[22]

The question that arises is: can a river transformed by dams into a linear chain of lakes, often completely detached from one another, even be called a river? The goal at the very least is to isolate energy pods behind a dam embankment that, like a giant spigot, can be turned on and off to generate an electric pulse. Only when the dams fail, in, say, a massive flood, do we witness a river that has reclaimed its watershed.

SEWAGE, WASTE, IRRIGATION, AND DRINKING WATER

The purpose that state engineers and corporations assign to a river dictates in turn the optimum conformation best suited to that particular goal. It is not unusual for a river to be managed for dual purposes—for example, irrigation and hydroelectric power—but in a pinch, one hegemonic purpose will prevail.

Some purposes of river engineering are largely beneficial to humans and nonhumans alike. A most obvious example is river management to achieve drinkable water, a goal largely compatible with the needs of fish, birds, and other river species. A counterexample is chains of hydroelectric dams that disconnect the river and destroy adjacent wetlands—a disaster for migrating fish and most other riverine species. Rivers designed primarily for irrigation (usually for a dry-season crop) require impounding water in reservoirs that can be dispersed to crop fields during the dry season. In this regard, they have a comparable effect to that of hydroelectric dams, depriving the wetlands and floodplains downstream of sufficient water to maintain the varied habitat that sustains most of the watershed's biodiversity.

The age-old function of rivers in removing human waste downstream to the sea presents dangers and risks that are relatively straightforward to understand. Upstream communities have the upper hand: having no one upstream of them, they can pass any water they pollute downstream, contaminating the water for the flora, fauna, and humans who dwell there. Midstream communities, of course, both receive polluted water from upstream and contribute their own waste matter, counting on the river to send it further downstream. Those most affected, obviously, are the human and nonhuman inhabitants of the delta and its distributaries.

This sketch of the transportation of waste downstream is, however, far too abstract to do justice to the actual process. The impression that the river is a steady conveyor belt is wildly off the mark. First, the movement of wastewater is highly irregular: if the banks of the river have been deforested and replaced with annual crops, if dikes and levees confine the channel, if drainage is extensive, if the gradient is relatively steep, and if the monsoon rains are strong, the volume and speed of each flow will likely scour sediment from the riverbed. If, on the other hand, the gradient is minimal, and the dry season features little rainfall and hence a current of lower volume and speed, the river is likely to slow to the point where a great deal of sediment, including toxic materials, is deposited on the riverbed. The river not only moves waste; it is also a massive earth-moving machine, conveying impulses of soil, sand, clay, pebbles, shells, and even boulders downstream.

What is utterly distinctive, however, and an effect of the industrial revolution, is the exponential explosion of waste matter that is toxic to most riverine creatures. Pollutants, largely nondegradable, unlike raw sewage, accumulate in the organs and tissues of fish, birds, and insects and disproportionately affect life downstream. Industrial waste, especially concentrated in the Monywa and Sagaing regions of the Ayeyarwady; pesticides and fertilizers from cropped areas in the watershed; and that universal plague of most watercourses, plastics in millions of forms increasingly devastate the life of the river. Singly and collectively, these pollutants account for the observed decline in both the volume and diversity of fish, birds, and mammals in the watershed. One suspects that they have had much the same effect on reptiles, amphibians, crustaceans, and insects, although no census figures are available.

The River's Woes Are Iatrogenic

Whenever possible, I try to avoid unnecessary technical terms and neologisms in the interest of clarity. I make an exception in this section by introducing the medical term *iatrogenic*. In its simplest form, an iatrogenic illness is one caused by previous treatment or nosocomial infections contracted in hospitals or clinics, such as the bacterial infection streptococcus, which is highly resistant to many antibiotics.

The reason for introducing this term from a different realm is simply because, as we shall see, most of the disasters of rivers with which we grapple today are the results of prior efforts to discipline and domesticate rivers for the benefit of *Homo sapiens* and their nation-states. Our efforts to sculpt the river for our purposes, and above all to simplify the river and the watershed altogether, are the major causes of river illness: from the erasure of wetlands to deforestation to the loss of habitat necessary for biodiversity to the decline or extinction of whole species of fish, plants, and riverine life-forms. Many of the effects of our interventions into river hydrology could not have been foreseen at the time they were made. Others were understood, but did not impede the practices designed to extract maximum value from their use and exploitation.

One example is summarized in a 1928 report by the Orissa (a province of eastern India) Flood Committee:

> We have come to the conclusion that the problem which has arisen in Orissa is due in the main to the efforts which have been made towards its protection. Every square mile of country from which spill water is excluded means the intensification of floods elsewhere; every embankment means the heading up of water on someone else's land. Orissa is a deltaic country and in such a country floods are inevitable; they are Nature's method of creating new land and it is useless to attempt to thwart her in her working. The problem in Orissa is

not how to prevent floods, but how to pass them as quickly as possible to the sea. And the solution lies in removing all obstacles which militate against this result. . . . To continue as at present is merely to pile up a debt which will have to be paid, in distress and calamity, at the end.[23]

As a first step in pursuing this analogy, it helps to elaborate on the characteristics of iatrogenic illness in medicine. Roughly 70 percent of hospitalizations in the United States are the direct or indirect result of prior treatment. The side effects of chemotherapy, radiation, and surgical procedures are obvious examples, and medical experts have endeavored to mitigate these effects as part of their treatment regimen. More troubling, however, is the worldwide phenomenon of antibiotic resistance. A medical miracle that has saved millions of lives since 1945, antibiotics were widely used not only to treat bacterial infections like staphylococcus, but in profligate ways in industrial livestock rearing: chickens, pigs, and beef and dairy cattle, to name the most notable cases. The two reasons for such profligate use bring us back to the simplification and crowding of modern industrial livestock production—both aimed at maximum profit for the least cost. The species are of a relatively uniform genetic makeup—selected for rapid growth, the quality of their meat, and the economy of raising them—and the vast majority of these animals are kept in small, confined spaces. Such simplification and crowding create a perfect epidemiological storm in waiting. The bacteria that are attracted to these animals can rampage at exponential speed, killing the entire flock or herd. The vulnerability of such one-commodity industrial production is analogous to the German scientific forest; replanted with one species, all the same age, it was exposed to an epidemic of rusts and other fungal infections that threatened the entire forest.

Bacteria, however, are both adaptable and more diverse. If, as is often the case, a particular variant of the bacteria confers resistance to the antibiotic targeting it, that resistance strain will proliferate, often at an exponential speed, displacing the bacteria that are not resistant.[24] The resistant strain reigns supreme unless and until the creators of antibiotic drugs can stymie it in its tracks. Subsequently, the process continues as before. Strains of bacteria that have by mutation, gene exchange, or hybridization become resistant have to be targeted by a new drug tailored to its unique vulnerability. The shorter the life cycle of the pathogen, the more quickly a new resistant strain is likely to appear.[25]

The focus thus far has been on iatrogenic effects among animals. Much the same logic, however, applies to fungal diseases in plants. Here, as in general, environment makes all the difference. The more simplified the landscape and soil, the narrower the genetic diversity of the cultivar, the greater the crowding of the plant in question, and the more severe the selective pressure, the more likely that strains of fungi resistant to the fungicide and fumigants to which they have been exposed will arise.[26]

A sketch of the most recent and consequential phase of river engineering would be misleading if it did not emphasize the central role of the state. As most premodern cereal-growing states were sited on the floodplain, in an era of technological simplicity the most vital resource of the state was the size of its mobilizable, productive population. That population, including slaves, provided accessible grain that could be taxed and stored, a pool of conscription in the case of conflict, and a labor force that could be deployed for public works. Holding onto that population—and whenever possible augmenting it—in the face of crop failures, epidemics of domesticated livestock and humans, ruinous levels of taxation, and the leakage of runaways to the ungoverned frontier was the most vital element of statecraft. This, in turn, meant using whatever means were available to ensure the availability of the water necessary for cropping (by building canals, dug tanks, weirs, and dams) and to minimize the danger of a catastrophic flood (by building dikes, levees, bunds, weirs, and drainage canals).[27]

Anthropocentrism

For an author trying, often in vain, to curb his anthropocentrism, what is striking is the degree to which the iatrogenic effects of river engineering have been centered on *Homo sapiens* to the relative neglect of all other living organisms. This emphasis is, to be sure, in keeping with the literature on iatrogenic illnesses, which concentrates on the impact of medical procedures on the patient being treated and only rarely on the wider effects on humans, let alone nonhuman life. Humans have benefited to some degree by river engineering (irrigation water, draining of agricultural lands, travel and navigation, and less frequent floods). They have also suffered. Taken together, the deforestation of the watershed, the replacement of old-growth, diverse forests by annual shallowly rooted crops or pastureland, and the straightening of the river and elimination of meanders have contributed to what has been termed the Great Drying. The amount of erosion and sediment entering the river, especially in the wet monsoon season, increases enormously, and the speed of the flood pulse current is greatly increased. If we add dikes and levees to the equation, the flood pulse is confined to the main channel, so, unless the speed of the current is always sufficient to scour the riverbed, the sediment is denied access to the floodplain, the wetlands, and the biodiverse in-between niche, meaning that most of it will be deposited on the riverbed, maybe actually raising the riverbed higher than the adjacent floodplain.

One astonishing example of this occurred near Kaifeng, where the Yellow River enters the low-gradient floodplain, as described in chapter 1. This topsy-turvy effect of human intervention is not rare (the lower Mississippi is another example), but the Chinese case is the most dramatic.

The imperative of holding productive populations in place and defending them against floods has been aptly described as political lock-in. As a description, it applies to modern democratic states as much as to ancient kingdoms. Once a tax-paying, growing population is settled beside a river on a fertile floodplain, the state has a vital interest in defending it—and the services it provides. In democratic settings, a voting population and its representatives are insistent that the state defend their land, their houses, and their businesses, whatever the cost, and they often have the political clout to prevail in contemporary politics. The ubiquitous cost-benefit analysis now required for all infrastructure projects outlines the economic benefits—the hypothetical productivity of farms, shops, businesses, and industries that might be built if you promise the dikes will prevent flooding.[28]

One obvious result of walling off cropland from the flood pulse was to deprive the soil of the rich nutrients that had made it productive in the first place. For farmers in many locations, the total loss of nutrients was averted by poking a small hole in the dike to let in the precious brown water. A second result of walling off the water was to accelerate the Great Drying, which in turn led to the subsidence of the floodplain landscape. We have noted this process on a much smaller scale in the horseshoe dikes around villages in the Ayeyarwady Delta. In terms of landscape dynamics, the two processes reinforced one another. As sediment accumulated outside the horseshoe, the plain where crops were grown sank. This growing gap portended more severe floods.

The effort to prevent floods entirely has strong parallels with the effort to prevent all bacterial infections. Both are stellar, if sobering, examples of iatrogenic effects. The profligate use of antibiotics in livestock feed, like its use on humans for a host of quite minor infections, creates a force field of selection that favors those bacteria that have the capacity to withstand the antibiotics. As the population of resistant bacteria (candida) replaces nonresistant strains, it sets off a sometimes frenzied search for a new antibacterial agent that can eliminate it. The competition is rather like an arms race.

Flood prevention, minus the genetic component, follows much the same path. In the attempt to prevent virtually all floods, weirs, levees, and dikes are erected to confine the river to its channel. Depending on the extent of prior deforestation, erosion, and drainage, sediment is likely to accumulate on the riverbed, often raising its level above that of the surrounding floodplain. These barriers do indeed make floods less frequent by providing more space for the river to swell at high water before it overtops the levees and dikes. When, however, levees and dikes fail, the flow they unleash is likely to be far more destructive. *The aim of preventing all floods comes at the cost of laying the groundwork for more catastrophic floods.* Epic floods along the flood-protected Mississippi watershed in 1927 and 1993 wreaked unheard-of damage on towns and villages, croplands and factories.[29] A little-noted iatrogenic effect exacerbated their duration and destruction. Much of the floodwater backed up behind those sections of the dikes and levees, still blocking the water from reentering the main channel, so it drained and evaporated only slowly.

Floods and fire are entirely natural phenomena. They predate by millennia the appearance of *Homo sapiens* on the planet. Even today, the vast majority of floods and fires are not caused by humans. And they can be useful. The use of floodwater was vital in detoxifying many foods, irrigating, nourishing crops and livestock, and turning waterwheels to mill grain. Similarly, fire was humankind's first and most powerful tool to clear land, cook, and provide warmth, not to mention keep predators at bay. Not surprisingly, the impulse to eliminate altogether these two vital resources backfired. The suppression of all fires for much of the past eighty years has led to the accumulation of combustible material, with the result that although there are fewer fires, when they do occur, they are, like the floods of the modern era, far more catastrophic.

What about the Rest of Living Organisms?

The term *iatrogenic* serves an important purpose in our discussion. It reveals how attempts to treat and cure illnesses often have inadvertent side effects that may not only negate the cure but could in fact create even more serious illnesses that are harder to treat. The concept helps to highlight the unforeseen consequences of humankind's efforts to prevent floods and to extinguish all wildfires as quickly as possible. For our purposes, however, it has one fatal flaw. It focuses resolutely on the human consequences, overlooking the effects on almost all other forms of life.

To do so is, of course, to ignore the overwhelming majority and astounding variety of life-forms that contribute to the base of the nutritional pyramid that sustains the mammalian and human world. Alas, as Thom Van Dooren puts it: "There is a public prejudice against invertebrates."[30] The invertebrate world represents by some calculations between 95 and 97 percent of all living creatures on the planet. Among them are beetles, ants, flies, butterflies, snails, molluscs, and crustaceans such as shrimp, crabs, and crayfish (to mention but a few). Each of the river's tributaries flowing through its distinct soils and landscape contributes its own distinctive mix of invertebrate and vertebrate species to the main river. Regrettably, and for reasons that are by now familiar, the data we have on the health and population of invertebrates, particularly insects and aquatic plants, are woefully incomplete and lack any deep temporal dimension. The most striking exceptions to this pattern of selective ignorance are turtles and frogs, many of which are in immediate danger of extinction both in Burma and worldwide. In a series of observations reminiscent of Rachel Carson's attention to the loss of songbirds, some evidence has suggested that we are losing insect populations at an unprecedented rate of 2 percent a year, while a German study suggested a 75 percent decline by weight in insect populations over twenty-seven years. Talk has arisen of an insect apocalypse.[31] If there's even a modicum of truth to these observations, it serves to redirect our attention to the species at the bottom strata of the food chain, including the vital pollinators.

As *Homo sapiens* built their societies, the consequences of their transformation of nature involved a form of ecocide as cultivators cleared land, built dikes and drainage canals, and, simply by displacement, radically reduced biodiversity. Nowhere is this as pronounced as in the vulnerable freshwater of the riverine environment. The loss of freshwater species is twice that of the species loss in brackish and saltwater. Virtually all these losses in biodiversity are anthropogenic effects.[32]

Habitat destruction is the dominant cause of the loss of biodiversity. In most cases, it can be attributed to the appropriation (theft?) of water from the habitat of nonhumans for human purposes. Worldwide, humans have come to divert more than half of the Earth's surface water for purposes of their own.[33] One result is that 40 percent of freshwater aquatic plants are threatened by extinction. There are at least two different ways of putting this process in perspective. The first is to acknowledge that from the late Neolithic onward, the expansion of human settlement and cultivation has required as an essential condition the extirpation of forests, the drainage of wetlands, the blockage of the floodplain, and the creation of an erosion-prone riverbank. Although the utter contradiction between the landscape required for human expansion and that needed by amphibians, insects, aquatic plants, and fish accustomed to feeding on the floodplain was obvious analytically, the contradiction was imperceptible, as was its impact on nonhuman populations, because the human population and the area of cultivated land increased slowly. The second perspective is to regard *Homo sapiens* as an invasive species, like beavers in the sense that they can modify the environment in their favor.

Tropical waterways are, in general, a good deal richer than temperate waterways in terms of the biodiversity of both flora and fauna. Humans as invasives both interrupt and destroy their complex productivity. First, they deprive nonhuman species of vital water they need by drainage, irrigation by dams, and dikes and levees keeping water from the floodplain. Second, they transform the landscape from a complex environment with marshes, mangroves, and a cornucopia of flora and fauna into a radically less biodiverse landscape of annual crops. Another result is a fragmented ecology in which connectivity is vastly reduced, and in many cases, the in-between, which joins dry and wet areas, is eliminated.

The Dry Zone and Burma itself count for 70 percent of the total surface water extracted from the Ayeyarwady, while the Mu River Valley to the west extracts 48 percent of its volume during the dry season. In the modern era, the water that remains is likely to be polluted with fertilizer, runoff, herbicides, insecticides, sewage, industrial waste, and plastic debris, which accumulate downstream from Monywa and Sagaing on the way to the delta. The flora and fauna that survive this assault are increasingly endangered.

A Soft or Hard Path for River Engineering

In my passion for orienting this book toward the nonhuman creatures so often ignored in the anthropocentric studies of river basins, I may have overlooked that humans are also riverine mammals with their own modest claim to a riverine livelihood—short of totally remaking the river to do our human bidding: for example, as a navigation channel, a series of irrigation ponds and reservoirs, a sewage pipe, or a series of hydroelectric dams. That form of hard-path river engineering involves extensive dredging, diking, and straightening. In short, it involves transforming the river into a disciplined, predictable, regulated, one-commodity machine. As we have seen, hard-path engineering of this sort, especially engineering to prevent floods, while likely to result in fewer small floods, raises the possibility of larger, more catastrophic floods.[34]

The principle behind the hard-path engineering of rivers is that water that falls, say, without turning a turbine or irrigating a crop is that much water wasted. This view, however, is totally myopic. The river is a living hydrological community that helps spawn, feed, and shelter large numbers of flora and fauna, from algae to insects to dolphins. The life-sustaining watershed system, if allowed to move as it will, is more productive of life and biodiversity than virtually any other natural system. Any intervention by hard-path engineering to maximize the return from a river is likely to damage its life-giving properties, making it far less productive over the long term. Most of the subsidies the river provides to biodiversity are invisible through the lens of cost-benefit analyses. The hubris embedded in cost-benefit analyses and their cousin ecosystem services is nothing short of staggering.[35] Given our wealth of ignorance about the environment and interspecies connections, it seems presumptuous to assume that hard-path engineers know more than the river.

What, then, is soft-path river engineering? One way to visualize its spirit is to liken it to a footpath in the forest. Suppose a tree falls across a footpath. A soft-path response would, in a Taoist spirit, be simply to divert the trail around the fallen tree. A more interventionist response would be to remove the tree and restore the original route. A still more interventionist response might be to straighten and pave the path to embed it more permanently in the landscape. The true high-modernist step would, of course, be to create a superhighway that removes the landscape and bulldozes straight through all obstacles in the topography. Soft-path engineering has the singular advantage of intellectual modesty with respect to what we actually know about river movement and its environmental effects. In contrast to hard-path engineering, soft-path engineering accepts variability in the river's movement as valuable until proven otherwise. Meanders, backwaters, ephemeral wetlands, braids and channels, swamps—all anathema to hard-path engineering—are presumed by soft-path engineers to be biotically important.

The report by the World Commission on Dams in 2000 was perhaps the most decisive moment to date in a retreat from hard-path engineering.[36] The commission noted the prejudice in favor of large dams, despite their social costs, their largely unanticipated environmental costs, and their failure to provide the projected benefits in almost all cases. They advised that dams not over fifteen meters in height were more likely to be beneficial. Since then, there has been a flurry of initiatives to favor dams that minimize intervention into the hydrology: low-head dams, run-of-the-river dams, or single-anchor turbines that use flow power to generate electricity for a single village.

Efforts to restore wetlands and floodplains by environmental agencies and nongovernmental organizations are ubiquitous. Their initiatives to remake hard-path landscapes where they have been implemented have shown striking results in terms of marked increases in biodiversity. Such restoration efforts are in keeping with indigenous concepts of rivers as living beings and modern attempts to legally codify the personhood of rivers. Their collective impact, however, is comparatively trivial in a landscape where farmers, homeowners, industries, and their representatives have become dependent on short-term benefits for their security. They, together with geoengineers who are concocting huge interventions in the natural world, are, sadly, likely to prevail—perhaps the most demoralizing sign of even soft-path engineering is that it is resolutely focused on the needs of *Homo sapiens* for drinking and irrigation water as well as energy and flood protection. The main interest nonhuman species share unambiguously with humans with respect to a river is unpolluted water, which is, not incidentally, key to the supply of the fish, shrimp, and molluscs that are at the center of the Burmese diet.

As Hegel famously expressed it, "The owl of Minerva flies only at dusk." As people began to understand how environmentally disastrous their redesign of the woodlands had become, the Germans, coiners of the term *Waldsterben,* or forest death, took simple steps to create a more biodiverse forest. In a similar fashion, the Dutch, previously known for their feats of hydraulic engineering to recruit and reclaim land from the sea, have, in the last decade, initiated the movement Make Room for the River, a recognition that they pushed land reclamation to a self-defeating and environmentally destructive degree. The pioneers of scientific forestry and of land reclamation from wetlands and the sea are the first to encounter the unexpected costs and limits of their techniques. Perhaps we can find a glimmer of optimism in their newfound modesty. If we heed the voice of the Ayeyarwady River speaking on behalf of the more-than-human world, we will have taken the first step on a more promising path.

NOTES

Introduction

1. I distinguish below between a "thick" and a "thin" Anthropocene. The former represents the conventional use of the term *Anthropocene*, the onset of which is in dispute but which refers to that epoch in which human activity comes to dominate environmental change. The "thin" Anthropocene, by contrast, is my term for an earlier period beginning roughly with the beginnings of agriculture and lasting several millennia, until roughly the beginning of the industrial revolution.

2. Neil Shubin, *Your Inner Fish: A Journey into the 3.5-Billion-Year History of the Human Body* (New York: Vintage, 2009).

3. A minority (lunatic?) fringe with, however, the hegemonic power to have its way. One solution to the asymmetry of power, one exercised in principle by environmentalists, is to insist on nominating human proxies empowered (as are guardians and conservators) to act on behalf of natural entities that have no obvious means of articulating their own interests. This is the path suggested by Christopher Stone and Garrett Hardin in their classic *Should Trees Have Standing? Toward Legal Rights for Natural Objects* (Los Altos, CA: W. Kaufman, 1974), 18.

4. George Perkins Marsh, *Man and Nature; or, Physical Geography as Modified by Human Action* (New York: Scribner, 1864).

CHAPTER ONE Rivers

1. Antoine Kremer, "Microevolution of European Temperate Oaks in Response to Environmental Changes," *Comptes rendus biologies* 339, nos. 7–8 (July–August 2016): 263–67, https://doi.org/10.1016/j.crvi.2016.04.014. For beeches, see Donatella Magri et al., "A New Scenario for the Quaternary History of European Beech Populations: Paleobotanical Evidence and Genetic Consequences," *New Phytologist* 171 (2006): 199–221, https://doi.org/10.1111/j.1469-8137.2006.01740.x.

2. When, in our lifetime, these certainties become dislodged, as in the case of climate change and its perilous implications for our future as a species, we are forcibly driven to a deeper historical inquiry than is customary. Without the growing catastrophes of sea-level rise, floods, fires, droughts, and so on to prod us, one doubts whether the concept of the Anthropocene would have become the subject of such an explosion of research.

3. Dilip da Cunha, *The Invention of Rivers: Alexander's Eye and Ganga's Descent* (Philadelphia: University of Pennsylvania Press, 2018).

4. It is not uncommon for a river to reclaim an older channel if it is now the best gravitational path to the sea.

5. Here I rely on the deeply researched and technologically sophisticated study by Ruth Mostern, *The Yellow River: A Natural and Unnatural History* (New Haven: Yale University Press, 2021). As a deep history of a single major river, it is in my view unsurpassed. Much the same process is observable in many flat river deltas. The Mississippi Delta is a striking case in point. For the past seven thousand years "it has swung back and forth a swath of coast nearly 420 kilometers wide" as it clogged one path to the Gulf of Mexico after another. It would already have moved to the steeper Atchafalaya River course to the west were it not for the massive engineering interventions of the U.S. Army Corps of Engineers. See Ellen Wohl, *A World of Rivers: Environmental Change on Ten of the World's Great Rivers* (Chicago: University of Chicago Press, 2010), 219, and the first section of John McPhee's *The Control of Nature* (Waterville, ME: Thorndike, 1999).

6. Mark Elvin, *The Retreat of the Elephants: An Environmental History of China* (New Haven: Yale University Press, 2006), 123–24.

7. David A. Pietz, *The Yellow River: The Problem of Water in Modern China* (Cambridge, MA: Harvard University Press, 2015), 19.

8. Another, more prudent traditional technique for "scouting the channel" on downstream voyages was to anchor and pay out a rock-laden raft with a draft slightly more than the mother ship. If, following the current, it did not run aground, the ship could then safely proceed. If it did run aground, it was retrieved and paid out again to the next most promising channel.

9. Theodor Schwenk, *Sensitive Chaos: The Creation of Flowing Forms in Water and Air*, translated by Olive Whicher and Johanna Wrigley, illustrations by Walther Roggenkamp (London: Rudolf Steiner, 1965).

10. This is, of course, true on a larger scale, as embodied in the iconic "hydrological cycle" that links sunlight and evaporation to cloud formation and rainfall and thence to wetlands, rivers, aquifers, and groundwater.

11. And when they did, at least in the Northern Hemisphere, millions of beavers had already amplified the process of slowing down rivers by building their own dams. See Ben Goldfarb, *Eager: The Surprising, Secret Life of Beavers and Why They Matter* (White River Junction, VT: Chelsea Green, 2018); and Ellen Wohl, *Saving the Dammed: Why We Need Beaver-Modified Ecosystems* (Oxford: Oxford University Press, 2019).

12. In this category, we should include gradual processes such as glacial melt or the movement of tectonic plates that reach thresholds that suddenly unleash epochal events such as the birth or extinction of a river.

13. See the suggestive laboratory experiment that appears to show landscape "creep" in wave form in the absence of other forces. One suspects that it has to do with the planet's movement and the tidal force of the moon. Adam Mann, "The Hills Are Alive with the Flows of Physics," *New York Times*, June 24, 2021, https://www.nytimes.com/2021/06/24/science/hills-creep-lasers.html. See also Nakul Deshpande, David Furbish, et al., "The Perpetual Fragility of Creeping Hillslopes," *Nature Communications* 12, no. 3909 (2021), https://doi.org/10.1038/s41467-021-23979-z.

14. More recently, specialists monitoring flood risks in the United States have placed gauges in watercourses, on the basis of which they devise flood-risk maps to guide insurance companies, zoning authorities, would-be developers, and disaster-management agencies like FEMA. These techniques failed utterly in August 2021 when a flood in middle Tennessee killed more than twenty people. The problem, apparently, was that the monitoring data were out of date and were gathered exclusively along the main river channel, ignoring the smaller streams and tributaries that were the sites of most of the flooding. In retrospect, the risk seems to have been grossly underestimated. See Christopher Flavell, "How Government Decisions Left Tennessee Exposed to Deadly Flooding," *New York Times,* August 29, 2021, A14. The term *flash flood,* as near as I can make out, references not only a local and fast flood but, above all, one that was not anticipated in the models of the experts. See also P.C.D. Milly et al., "Stationarity Is Dead: Whither Water Management?" *Science* 319, no. 5863 (February 2008): 573–74, https://doi.org/10.1126/science.1151915.

15. George Everest, *Historical Records of the Survey of India,* vol. 4: *1830–1843,* ed. R. H. Phillimore (Dehra Dun: Surveyor General of India, 1958), 266.

16. Wohl, *A World of Rivers,* chapter 5.

CHAPTER TWO In Praise of Floods

1. Peter B. Bailey, "The Flood Pulse Advantage and the Restoration of River-Floodplain Systems," *Regulated Rivers: Research and Management* 6, no. 2 (April–June 1991): 75–86, https://doi.org/10.1002/rrr.3450060203. The quantification, to be specific, is measured here by fish yield (by weight!) "per unit of mean water area."

2. Wolfgang Junk, Peter B. Bailey, and R. E. Sparks, "The Flood Pulse Concept in River-Floodplain Systems," in *Proceedings of the International Large River Symposium,* ed. D. P. Dodge, *Canadian Special Publications of Fisheries and Aquatic Sciences* 106 (1989): 110–27.

3. Seth R. Reice, *The Silver Lining: The Benefit of Natural Disasters* (Princeton: Princeton University Press, 2001), 110.

4. Here I follow the useful description in Reice, *The Silver Lining,* 112–15.

5. Reice, *The Silver Lining*, 115. The contribution of flooding to terrestrial and aquatic productivity is also crucial to the biodiversity of that productivity.

6. Ellen Wohl, *A World of Rivers: Environmental Change on Ten of the World's Great Rivers* (Chicago: University of Chicago Press, 2010). On the anthropogenic breaking of "connectivity," see her deeply researched *Disconnected Rivers: Linking Rivers to Landscapes* (New Haven: Yale University Press, 2004).

7. Wohl, *A World of Rivers*, 120. We have concentrated on the nutrients available to bacteria at the base of the aquatic food chain and fish, the most studied because commercially valuable, near the top. The analysis could just as appropriately apply to amphibians such as toads, frogs, newts, and salamanders or to turtles, lizards, crocodiles, alligators, lung fish, mudskippers, and betta fish, not to mention plant, avian, and mammalian life. The analysis of rivers, in general, could be justly accused of a "pisco-centric" bias, reflecting the narrow concerns of *Homo sapiens*.

8. One might say that the transition zone favors generalists, not specialists. Once I was asked what sort of nonhuman I might desire to be if given the choice. After a moment's reflection I replied that I would like to be a loon or a merganser. Why? Because they can fly, walk on land, swim on the surface, and finally dive and stay underwater for long periods. They pay a price for their lack of specialization, but they can navigate in all four mediums.

9. It does, of course, happen naturally when migrating birds unwittingly carry fish eggs, when a massive flood temporarily joins two watersheds, and when *Homo sapiens* deliberately or inadvertently transfers a species to a new location. For this last, see James Prosek, *Eels: An Exploration, from New Zealand to the Sargasso, of the World's Most Mysterious Fish* (New York: HarperCollins, 2010); and Anders Halverson, *An Entirely Synthetic Fish: How Rainbow Trout Beguiled America and Overran the World* (New Haven: Yale University Press, 2010).

10. Similar conditions in other watersheds have stimulated comparable traits, an environmentally driven process called "convergent evolution."

11. Carson E. Jeffries, Jeff J. Opperman, and Peter B. Moyle, "Ephemeral Floodplain Habitats Provide Best Growth Conditions for

Juvenile Chinook Salmon in California Rivers," *Environmental Biology of Fishes* 83, no. 4 (2008): 449–58.

12. Mickey E. Heitmeyer, "The Improvement of Winter Floods to Mallards in the Mississippi Alluvial Valley," *Journal of Wildlife Management* 70, no. 1 (2006): 101–10.

13. Ian G. Baird, "Fishes and Feasts: The Importance of Seasonally Flooded Riverine Habitat for Mekong River Fish Feeding," *National Historical Bulletin of the Siam Society* 55, no. 1 (2007): 121–48. Baird also records as many as thirty-five different forest fruits as well as floodplain plants in the stomachs of netted fish.

14. I owe this account to the work of Abbie Gascho Landis, *Immersion: The Science and Mystery of Freshwater Mussels* (Washington, DC: Island, 2017). Her account is, in my view, something of a model for naturalists who write for nonspecialists.

15. Gascho Landis, *Immersion*, 70. For an account of mussel "radiation," or colonization across watersheds, Ivan N. Bolotov et al. emphasize the ancient high-water marks in bursts between twelve thousand and seven thousand years ago when watersheds were most connected. "Ancient River Inference Explains Exceptional Oriental Freshwater Mussel Radiations," *Scientific Reports* 7, no. 2135 (2017), https://doi.org/10.1038/s41598-017-02312-z.

16. The agricultural year has its own rhythm, of course, one dictated by the genetic clockwork of the most important crops, particularly at harvest time. It is only by comparison with hunting and foraging that it seems to involve more toil. Workers in both systems are tied more closely to natural rhythms than, say, the office or factory worker, whose movements are synchronized by the clock, the machine, and productivity metrics.

17. The region's nutritional richness was greatly enhanced by nearby marine resources underwritten by rich kelp beds along the coast throughout the Strait of Juan de Fuca. See the important book by Joshua L. Reid, *The Sea Is My Country: The Maritime World of the Makah* (New Haven: Yale University Press, 2015). The name Makah, bestowed on them by a neighboring group, translates as "generous with food."

18. The exceptions were the choke points in trade routes (such as a mountain pass or a narrow strait), the control of which could become a source of tribute in grain or other valuables that could, in turn, sup-

port a substantial population. Melaka, astride its eponymous strait, is a striking case in point.

19. Thomas K. Park, "Early Trends toward Class Stratification: Chaos, Common Property, and Flood Recession Agriculture," *American Anthropologist* 94:90–117, https://doi.org/10.1525/aa.1992.94.1.02a00060.

20. I am indebted here to the comprehensive work of the late Janice Stargardt, *The Ancient Pyu of Burma: Early Pyu Cities in a Man-made Landscape*, vol. 1 (Cambridge: PACSEA, 1990).

21. The symbolic claims of monarchs, on the other hand, to guarantee good harvests through their divine intervention are common. The annual ceremony of a ritual plowing by the reigning monarch is widespread in Southeast Asia. See, for example, Pe Maung Tin and G. H. Luce, trans., *The Glass Palace Chronicle of the Kings of Burma* (London: Oxford University Press, 1923), 51. For traditional agriculture in the "Dry-Zone," see Michael Aung-Thwin, *Irrigation in the Heartland of Burma: Foundations of the Pre-colonial Burmese State* (DeKalb: Northern Illinois University, Center for Southeast Asian Studies, 1990).

22. This border was often a point of ethnic, cultural, and linguistic demarcation, indicating, in the parochial view of lowlanders, the line between the "civilized" and the "barbarians." For an elaboration of this theme, see James C. Scott, *Against the Grain: A Deep History of the Earliest States* (New Haven: Yale University Press, 2017), chapter 7.

23. For this and subsequent comparisons, see Scott, *Against the Grain*, 266n15.

24. Log rafts were ubiquitous on the Mississippi until the twentieth century. Readers of Mark Twain's novel *Huckleberry Finn* will recall that Huck and Jim, an escaping slave, rode down the river on just such a raft.

25. These rafters on the Ayeyarwady have to maneuver huge and unwieldy vessels. They therefore have to choose a time when the river is not in spate and when it is not so low that they risk running aground frequently. Although, like their Mississippi counterparts, they have long oars at the stern that give them a bit of maneuverability, the fact that they can move no faster than the current dictates means that they lack "steerage." There are also cases where the vessel is the cargo as well. Before roads were available, farmers in upland central Pennsylvania had but one chance annually to bring their commodities (for example, but-

ter, cheese, whiskey, maple syrup, furs, and potash) to market. They built rafts to carry these products downstream to valley market towns in late spring when stream water was sufficiently high. They carefully selected the wood to construct each raft because it too was part of the cargo. It was dismantled and sold at the end of the one-way trip. When I stayed in my cabin in Pennsylvania I frequently fished off a huge boulder locally known as the Butter Rock. According to legend, one such raft on the way to market had broken up on the boulder and so much butter from the broken casks remained on the rock that it reeked of rancid butter for the remainder of the spring and summer.

26. I oversimplify here. Most premodern transportation of goods involved complex segments of travel by water, pack animals, and human porters. The law of friction still applies. Lewis and Clark were looking for a northwest passage by *water*.

CHAPTER THREE Agriculture and Rivers

1. I am indebted to the insight and analysis of George Perkins Marsh, whose *Man and Nature; or, Physical Geography as Modified by Human Action* (New York: Scribner, 1864) laid out a convincing case for the degree to which preindustrial humankind had altered the landscape, particularly river basins.

2. Though perhaps not so firewood-intensive as smelting, the slaking of limestone to produce plaster/mortar for interior surfaces and walls was significant. Evidence of its use is found in ancient Egypt as early as 6000 BCE.

3. Robert T. Deacon, "Deforestation and Ownership: Evidence from Historical Accounts and Contemporary Data," *Land Economics* 75, no. 3 (1999): 341–59, https://doi.org/10.2307/3147182.

4. Steven Mithen, *After the Ice: A Global Human History, 20,000–5000 BC* (Cambridge, MA: Harvard University Press, 2003).

5. Cited in Ruth Mostern, *The Yellow River: A Natural and Unnatural History* (New Haven: Yale University Press, 2021), 85. The passage comes from the second century CE but refers to an earlier legendary period when Yu the Great is said to have tamed the waters in the service of state power.

NOTES TO PAGES 74–92

6. The logic here is analogous to the effect latitude has on species diversity. At higher latitudes the environment becomes more constraining compared to that at the equator. The result is fewer species—though often in large numbers—while at the equator there are far more species—though often fewer of each. The contrast between birch and fir forests at cold latitudes versus the abundance of tree species in the tropical rainforest is a case in point.

7. Ellen Wohl, *Saving the Dammed: Why We Need Beaver-Modified Ecosystems* (Oxford: Oxford University Press, 2019), 6.

8. See the brief and mesmerizing summary of this process in Marsh, *Man and Nature*, 269–70n18. He is summarizing the more elaborate description of Wijnand C. H. Staring, *De Boden van Nederland I* (Haarlem: A. C. Kruseman, 1856), 36–43.

INTERLUDE

1. Maung Maung Oo and Naing Tun Lin are essential collaborators in these accounts. I have edited their narratives for clarity and concision. Their account of *nats* is a valuable reminder that the Burmese have a far more capacious understanding of the agency of nature and, in particular, the history and whims of the river that are beyond the efforts of hydrologists to tame it. For the humans living by the river, *nats* are an integral part of its distinctive history, biography, and personality, its threats and its benefits, and its potential role as a protective intermediary to ensure their safety.

2. My knowledge of the ecology of tropical watercourses comes thanks to David Dudgeon's collection: *Tropical Stream Ecology* (Amsterdam: Academic, 2008).

3. Dudgeon, *Tropical Stream Ecology*, 33. It is claimed as well that the number of species that migrate within the watercourse over their life cycle (for example, shrimp) are greater in tropical rivers.

4. Together, the Chindwin and the Ayeyarwady approximate the other great river of Burma, the Thanlwin (also known as the Salween and, in China, the Nu Jiang), which originates in the Himalayan highlands and flows through southwestern China and eastern Burma before emptying into the Andaman Sea at Mawlamyine in lower Burma. It

flows through mostly minority ethnic areas and through long gorges that sharply limit the larger floodplains so conducive to the populous settlements and riverine trade that otherwise characterize the Ayeyarwady. It is navigable for a mere ninety kilometers upstream from its estuary, further constricting trade and settlement. The Thanlwin merits more biographers than it has thus far attracted.

5. These minor tributaries flow, for the most part, through the rain-starved Dry Zone, though they were vital for the irrigated land that constituted the core of Burman (and Pyu) civilization. The data and images on which this description of the river is based derive, unless otherwise specified, from the comprehensive works of Benoit Ivars and Charles-Robin Gruel et al., on whose detailed research findings and reports I have relied. Benoit Ivars, "More Than Rice: A Contribution to the Ethnographic History on Resource and Frontier-Making in the Ayeyarwady Delta" (PhD diss., University of Cologne, 2022); Charles-Robin Gruel, "Sédimentation et érosion dans le Delta de l'Ayeyarwady: Focus sur trois archipels situés sur le cours principal du fleuve" (report prepared for the World Wildlife Federation, June 2018); Charles-Robin Gruel, Jean-Paul Bravard, and Yanni Gunnell, *Geomorphology of the Ayeyarwady River, Myanmar: A Survey Based on Rapid Assessment Methods* (Washington, DC: World Wildlife Fund, 2016); Ayeyarwady Integrated River Basin Management (AIRBM) Project, *Ayeyarwady SOBA 2017: Synthesis Report, State of the Basin Assessment*, vol. 1 (Yangon: Hydro-Informatics Centre, 2017).

6. In those sections of the Ayeyarwady that are, owing to highly friable soils, incised in the landscape, the monsoon rain pulse floods the flatland before the river itself overtops its banks. The result is that the flooded flatlands block a good deal of the sediment from dispersing. The surrounding landscape in this case may add as much sediment *to* the river as it receives *from* it. Such sediment will then be dispersed anywhere downstream where the river is not deeply incised.

7. Much of my initial understanding of the geological and hydrological history of the Ayeyarwady is owed to my exceptional research assistant Michael Lebwohl's report on such matters in 2016.

8. This is not uncommon in exceptional floods. See the stunning example depicted in William Faulkner's novella *Old Man*, where two convicts rowing a small boat in the epic 1927 Mississippi flood are attempting to rescue people stranded on levees. They are astonished to

find themselves swept upstream rather than downstream on the Yazoo River, a tributary of the Mississippi. "What [this convict] did know was it [the Yazoo] was now running backwards." William Faulkner, *The Faulkner Reader* (New York: Modern Library, 1959), 353–432.

9. There is a much smaller flood pulse in April, prior to the monsoon pulse, which is dependent on Himalayan glacial melt and snowmelt.

10. Movement is not solely predicated on sources of nutrition. Safety and suitable spawning sites also help explain migratory patterns.

11. A detailed and graphic account of a relatively recent large commercial flood-retreat fish-capture operation involving dozens of laborers and months of meticulous weir movement may be found in Benoit Ivars's fine account of U Lwin's practices in the delta: "More Than Rice."

12. Long before human and industrial wastes, the Ayeyarwady, like other rivers, also swept up in the flood pulse the detritus of the riverine landscape—dislodged silt and clay, plant wastes, dead invertebrates and vertebrates, mammalian and bird remnants, and the bacterial life dependent on that detritus.

13. The causes are a source of debate, with the volcanic activity at Mount Popa, the movement of tectonic plates, and sedimentation as the contending explanations.

14. Ivars, "More Than Rice," 227.

15. Ivars, "More than Rice," 209. One cannot but wonder about the micropolitics of this common-law procedure. The "main" channel quite possibly varies depending on the exact stage of the monsoon flooding or, for that matter, between the dry and wet seasons. In addition, it is all too common for the main channel to shift from year to year or at the flood peak, carving out a wholly new channel or jumping to a long-abandoned channel. When, precisely, the exercise is conducted seems decisive and there appears to be no evidence of a "replay."

CHAPTER FOUR Intervention

1. The apotheosis of improvement through engineering was the eugenics movement of the early twentieth century, in which selective breeding of plants and animals for desirable traits was extended to selection and breeding to achieve the perfection of *Homo sapiens*.

2. See, for example, Ian Hacking's illuminating *The Taming of Chance* (Cambridge: Cambridge University Press, 1990).

3. Only estimates are available because no reliable census has been conducted, given the political sensitivity around the proportion of the population that is of minority—not "Burman"—ethnicity, thought to be roughly 30 percent or more of the total. More recently, owing to the relatively lackluster performance of the economy and the vicious repression following the military coup of February 2021, the numbers fleeing the country have been substantial, if not precisely documented.

4. Ayeyarwady Integrated River Basin Management (AIRBM) Project, *Ayeyarwady SOBA 2017: Synthesis Report, State of the Basin Assessment*, vol. 1 (Yangon: Hydro-Informatics Centre, 2017). This is, to my knowledge, the most comprehensive and carefully documented source for almost any aspect of the river's hydrology, the river landscape, pollution, dams, and the changes over the last half century. I am deeply indebted to its research and judgments for my own knowledge of the overall river basin. Special thanks to Charles-Robin Gruel and Benoit Ivars for their part in the project and for bringing this and other associated reports to my attention.

5. One of the few victories of environmental politics in Burma was the successful campaign to stop the construction, already in its early stages in 2007, of a major hydroelectric dam at the confluence of the N'mai and the Mali Kha Rivers at the headwaters of the Ayeyarwady. Though noteworthy, this achievement has not had much effect on the building of dams on the river's major tributaries or on the Sittaung and the Salween/Thanlwin Rivers, despite local opposition.

6. The "official" human toll from Nargis was declared to be on the order of 140,000, although most knowledgeable observers consider the actual loss of life to be substantially higher. In 2015 I spoke with five men working in a delta rice mill, all of whom had lost their entire families, except for one worker's daughter, who survived because she was lashed to a tree.

CHAPTER FIVE Nonhuman Species

1. I puzzled long and hard over this move. It seemed both presumptuous and arrogant to even try to capture the agency of all creatures,

each of which experiences the world in vastly different ways. To reduce their ill-understood and unique sensibilities to spoken words on the page seemed both preposterous and, indeed, insulting. The classic of the genre and the founding text of biosemiotics was published in 1934 by Jakob von Uexküll: *Streifzüge durch die Umwelten von Tieren und Menschen* [A Foray into the Worlds of Animals and Humans] (Berlin: Verlag von Julius Springer). He was the inventor of the influential term *Umwelt*, understood as the perceptual world unique to each species. The example most cited from his work is the perceptual world of the tick.

At this point in our burgeoning understanding of the wondrous modes of cognition and communication among flora and fauna, it should be totally unnecessary to make the case for the "standing" of the more-than-human world. Because the Ayeyarwady dolphin plays a central role in what follows, the reader is referred to any one of the following accounts of the dolphin "world": Hal Whitehead and Luke Rendell, *The Cultural Lives of Whales and Dolphins* (Chicago: University of Chicago Press, 2014); Janet Mann, ed., *Deep Thinkers: Inside the Minds of Whales, Dolphins and Porpoises* (Chicago: University of Chicago Press, 2017); and Denise L. Hertzing and Christine M. Johnson, eds., *Dolphin Communication and Cognition: Past, Present, and Future* (Cambridge, MA: MIT Press, 2015).

A series of twenty books exploring various developments in biosemiotics, edited by Jesper Hoffmeyer, has been published by Springer. A more recent exploration of species-specific sensibilities is Ed Yong's *An Immense World: How Animal Senses Reveal the Hidden Realms around Us* (New York: Random House, 2022). A spate of books on trees and mycorrhizal connections offer worthy efforts to trace the contours of these various nonhuman ways of sensing the world. Each species represents a veritable encyclopedia of sensibilities that we are only beginning to glimpse. I am under no illusion that I have overcome these formidable obstacles. The choice finally was either to not give voice to nonhuman species at all or to attempt, however inadequately, to represent how changes in the river have impacted their life cycle. I apologize in advance for not doing justice to the fullness of their lifeworld.

2. If one doubted that the naming of the natural world was an imperial project, names such as this should dispel the illusion.

CHAPTER SIX Iatrogenic Effects

1. A partial exception are small states that control a vital choke point on overland or riverine trade routes, allowing them to extract foodstuffs and trade goods from the passing traffic. Timbuktu and Melaka are examples of such mini states in the early Common Era.

2. For a more elaborate exposition of this fragility, see chapter 3, "Zoonoses: A Perfect Epidemiological Storm," in James C. Scott, *Against the Grain: A Deep History of the Earliest States* (New Haven: Yale University Press, 2017).

3. Until recently, the process of conquering land from the sea was an integral and defining part of Dutch national identity. See, for example, Johan van Veen, *Dredge, Drain, Reclaim: The Art of a Nation* (The Hague: Martinus Nijhoff, 1955). More recently, Dutch environmental engineers have actually surrendered polders to the sea, and movements such as Room for the River (in this case, the Maas River) have questioned the wisdom of the earlier narrative.

4. Surely an undercount given the pace at which botanists are identifying previously unrecorded species.

5. Millet, another cereal grain, grown in the early Yellow River states, is an exception.

6. The drawback of legumes, I believe, is that they ripen sequentially, not simultaneously, so that when the taxman comes, only some of the fruits can be seized. The roots and tubers of the crop are relatively invisible and can, in fact, remain in the ground for two or even three years, to be dug up at the planter's leisure.

7. It was well known that planting a cultivar in virgin soil, or soil where that cultivar had never been planted, could lead to bumper crops. This was called the "honeymoon" period, during which the crop's "enemies" had been left behind and local parasites had yet to adapt to the new plant. This period, it is asserted, lasted roughly four hundred years when wheat from the Middle East came to Europe.

8. Well, not entirely. In virtually every large swamp, bog, fen, or marsh there was a population that lived and drew their subsistence from the wetland abundance here through fishing, hunting, gathering, and some cultivation. In many such areas, they resisted drainage

work with violence. Wetlands and marshes were also places of refuge for defeated rebels, runaway slaves, evaders of conscription and deserters, criminals, persecuted officials, and so on. See, for example, the Chinese classic: *The Water Margin Novel.*

9. One of those voices was George Perkins Marsh, *Man and Nature; or, Physical Geography as Modified by Human Action* (New York: Scribner, 1864). Although written in the mid-nineteenth century, Marsh's analysis is astoundingly prescient.

10. For a more elaborate account, see my *Seeing Like a State: How Certain Schemes to Improve the Human Condition Have Failed* (New Haven: Yale University Press, 1998), 11–22.

11. Jan Douwe van der Ploeg, "Potatoes and Knowledge," in *An Anthropological Critique of Development,* ed. Mark Hobart (London: Routledge, 1993), 209–27.

12. Vaclav Smil, "Eating Meat: Evolution, Patterns, Consequences," *Population and Development Review* 28, no. 4 (December 2003): 599–639.

13. I exclude, for the purposes of this discussion, domesticates that were chosen as draft animals—for example, donkeys, horses, llamas, and oxen.

14. It is worth noting that pigs are the exception when it comes to surviving in the wild. They have not only survived but thrived. In this respect they belong to a class of species, rather like bananas, daffodils, and salmon, that move back and forth.

15. Marc Elvin, *The Retreat of the Elephants: An Environmental History of China* (New Haven: Yale University Press, 2006), 137–38. He also built "set-back" dikes one kilometer or more from the primary dike as a second line of defense.

16. Meir Kohn, "The Cost of Transportation in Pre-industrial Europe" (Working Paper no. 01-02, Department of Economics, Dartmouth College, January 2001), https://ssrn.com/abstract=256600 or http://dx.doi.org/10.2139/ssrn.256600.

17. Mark Cioc, *The Rhine: An Eco-Biography, 1815–2000* (Seattle: University of Washington Press, 2006).

18. Though we have examined navigation through the lens of commerce, it is vital to understand that it facilitates social and cultural

connection. One key insight of Fernand Braudel's classic work, noted in chapter 2, was that places three hundred miles across easy water are likely to be closer in social, religious, linguistic, and customary terms than places a mere twenty to thirty miles away across difficult mountain terrain. The same is true for the Ayeyarwady River: Burmese Buddhism, the Burmese language, ethnic identity, dress, and spirit worship are to be found along a five-hundred-mile stretch of the navigable river, while thirty miles inland from its banks, one is likely to find a different language spoken, different ethnic identity, and different religion. The watershed unifies not only nonhuman riverine species but also its human inhabitants.

19. Quoted in Cioc, *The Rhine*.

20. The literature has been expertly summarized in Patrick McCully, *Silenced Rivers: The Ecology and Politics of Large Dams* (London: Zed Books, 1996).

21. Even prior to fossil fuel use, the fall line was key to early industry. It provided the power to turn grain mills, and millraces captured the inanimate power to turn looms and timber saws, all for only the cost of building the infrastructure. See Stefania Barca, *Enclosing Water: Nature and Political Economy in a Mediterranean Valley, 1796–1916* (Isle of Harris, UK: White House, 2010).

22. A very large dam at the headwaters was planned by the military with Chinese technical and financial assistance. In the early stages of construction, large-scale protests broke out, forcing the military to delay the work indefinitely. To some degree, this marked the beginning of the popular resistance to the military that has become a revolutionary civil war, in its third year as I write.

23. Bishop Svarup, D. G. Harris, and J. Shaw, *Report of the Orissa Flood Committee* (Patna: Superintendent Government Printing, 1928), 13.

24. Exponential growth has been used by the pioneering environmental economist Herman Daly in arguing for finite limits on both the growth of the total global output, which has increased more than thirty-six times in the twentieth century, and the nearly fourfold increase in the same period of the global population, from 1.6 billion to over 6 billion. He also maps the exponential growth in the case of

bacteria: "Our situation may have parallels to a well-known riddle. If the area of a Petri dish covered by bacteria doubles every hour, you inoculate the dish at noon on day one, and it is completely full at noon two days later (and therefore, the population crashes because it has exhausted its food source and inundated the Petri dish with waste), *when is the dish half full?* The answer of course is at 11am on the final day. At 9am, 7/8ths of the resources available for continued growth are still present. The question for humans is, how close is it to noon?" Herman E. Daly and Jason Farley, *Ecological Economics: Principles and Applications*, 2nd ed. (Washington, DC: Island, 2011), 112.

25. Paradoxically, the appearance of virulent resistant strains of viruses, bacteria, and fungi in hospitals is an effect of hygiene rather than its absence. Such effects are evident from what we now know about polio. If one were to take blood samples of, say, children from six years of age to fourteen in Mumbai, one would find that a vast majority of them carried antibodies to the polio virus, indicating that they had been infected when infants, when the infection is usually asymptomatic. Those who, thanks to hygiene, escaped infection when infants (it spreads via the fecal route) and who were first exposed as young adults were likely to suffer a more devastating illness. Those in very hygienic environments (preponderantly Westerners and urbanites) were spared a likely nonthreatening case in infancy at the cost of a potentially life-threatening infection post puberty. The same principle, as we shall see, is at work in flood control: the steps taken to prevent a small harm increase the risk of a far larger harm down the road.

26. Julie Guthman, *Wilted: Pathogens, Chemicals, and the Fragile Future of the Strawberry Industry* (Berkeley: University of California Press, 2019).

27. Most of the earlier literature on the state and water control greatly overstates the role of the state in irrigation, retainage, terracing, and flood control generally. See, for example, Karl August Wittfogel, *Oriental Despotism: A Comparative Study of Total Power* (New Haven: Yale University Press, 1957). Worldwide, most premodern systems of irrigation and flood control had been undertaken by one or more villages voluntarily mobilizing their labor to construct local waterworks to manage their local water resources.

Throughout this text I have used the terms *dike* and *levee* interchangeably. Technically, however, a dike is a barrier blocking water from inundating land that would normally be underwater, while a levee blocks water that might inundate a landscape that is normally dry but threatened (!) by floods or high tides.

28. Martin Doyle, *The Source: How Rivers Made America and How America Remade Its Rivers* (New York: Norton, 2018), 90–92. Doyle brilliantly describes the perverse process in which the cost-benefit analysis that is required for all such infrastructure (for example, dams and levees) is "gamed." The benefit side of the equation is predicated on the value and profits of farmland, neighborhoods, industrial parks, and even cities that could be built if floods were prevented. This became, then, the way to justify any flood-control project by virtue of the development that it made possible. Even then, subsequent analysis has shown that the projected benefits were inflated and the costs vastly underestimated.

29. Some of the most instructive analyses of the great floods of 1927 and 1993 are John M. Barry, *Rising Tide: The Great Mississippi Flood of 1927 and How It Changed America* (New York: Simon and Schuster, 1997); and Jeremy E. Gallowa Jr., "Learning from the Flood of 1993: Impact, Management Issues, and Areas for Research" (US-Italy Research Workshop on the Hydrometeorology, Impacts, and Management of Extreme Floods, Perugia, November 1995). Not to be overlooked is William Faulkner's novella about the flood: *Old Man* (New York: Signet Books, 1939).

30. Quoted in Kate Simpson, "Snail's Race to Extinction," review of *A World in a Shell: Snail Stories for a Time of Extinctions,* by Thom Van Dooren, *Times Literary Supplement,* May 19, 2023, 18.

31. Such findings also serve to question the temporal units we use in assessing changes in the Earth. Each generation, many environmentalists claim, takes its environment as the norm (for example, the prevalence of various insects and plants), having little or no awareness of, say, a long-term decline that would only register in a longer time horizon.

32. In a report presented in December 2023 to the Climate Summit, the International Union for the Conservation of Nature stated that

fully one-quarter of all freshwater fish were threatened with extinction. Fifty-seven percent of the threat was attributed to a combination of fertilizer and pesticide runoff, silt clogging rivers, land clearance and drainage, and human and industrial sewage. The remainder was due to water extraction by dams and irrigation, with overfishing and disease being relatively minor causes. Forty-two percent of amphibians were in decline, the most of any freshwater category. Catrin Einhorn, "A Quarter of Freshwater Fish Are at Risk of Extinction, a New Assessment Finds," *New York Times,* December 11, 2023, https://www.nytimes.com/2023/12/11/climate/climate-change-threatened-species-red-list.html.

33. In a devastating decision in 2013, the U.S. Supreme Court ruled that unless a body of water was permanently joined by a surface connection to a larger permanent body of water, it was not eligible for protection under wetland protection legislation. It was a decision based on a level of environmental illiteracy that boggles the mind.

34. Peter H. Gleick, "Water Management: Soft Water Paths," *Nature* 418, no. 373 (2002), https://doi.org/10.1038/418373a. For readers interested in forms of adaptive management that take into account environmental and institutional contexts, I recommend Lance H. Gunderson and C. S. Holling, eds., *Panarchy: Understanding Transformations in Human and Natural Systems* (Washington, DC: Island, 2012).

35. The technique designed to "smuggle" such ecological facts into cost-benefit analyses is known as "ecosystem services," in which products produced naturally by the river are, in effect, commodified and their value in monetary terms calculated. Some examples might be medicinal plants, construction material, fuel, and fodder. In practice these services are virtually all evaluated by what they provide to *Homo sapiens,* not what they contribute to other, nonhuman life.

36. World Commission on Dams, *Dams and Development: A New Framework for Decision-Making; The Report of the World Commission on Dams* (London: Earthscan, 2000).

ILLUSTRATION CREDITS

CHAPTER ONE Rivers

Movement of tectonic plates: Drawn by Bill Nelson based on USGS data.
Advances and retreats of glaciers: Adapted from data provided courtesy of Ayako Abe-Ouchi, University of Tokyo, Japan. Drawn by Bill Nelson.
European beech: Adapted from Donatella Magri et al., "A New Scenario for the Quaternary History of European Beech Populations: Paleobotanical Evidence and Genetic Consequences," *New Phytologist* 171 (2006): 199–221, figure 9, https://doi.org/10.1111/j.1469-8137.2006.01740.x. Drawn by Bill Nelson.
The Yellow River: Reproduced with permission from Ruth Mostern, *The Yellow River: A Natural and Unnatural History* (New Haven: Yale University Press, 2021), plate 1 and figure 3.18.
Mississippi River meander belt: Created by Harold Fisk, USGS.
Map of the Ayeyarwady River: Drawn by Bill Nelson.
Grounded vessel: Photo by author.
Map of water veins, Brenz River: Reproduced from Theodor Schwenk, *Sensitive Chaos: The Creation of Flowing Forms in Water and Air*, translated by Olive Whicher and Johanna Wrigley, illustrations by Walther Roggenkamp (London: Rudolf Steiner, 1965), 76, by permission of the original publisher, Verlag Freies Geistesleben.

Meanderness of a river; diagram of a river meander: Drawn by Bill Nelson.

Meanders of Nowitna River: Photo by Oliver Kurmis, CC BY-SA 2.0.

CHAPTER TWO In Praise of Floods

Examples of lateral shift of the Ayeyarwady River: NASA.

The flood pulse: Reproduced with permission from Peter B. Bayley, "Understanding Large River-Floodplain Ecosystems: Significant Economic Advantages and Increased Biodiversity and Stability Would Result from Restoration of Impaired Systems," *BioScience* 45, no. 3 (March 1995): figure 1, by permission of Oxford University Press.

The Dry Zone in Burma; the Ayeyarwady River: Drawn by Bill Nelson.

Raft of jars, 2010: Photos courtesy of Professor Soe Kyaw Thu.

CHAPTER THREE Agriculture and Rivers

Pattern of upstream deforestation: Originally published in James C. Scott, *Against the Grain: A Deep History of the Earliest States* (New Haven: Yale University Press, 2017), figure 14.

INTERLUDE

Major tributaries; ecological zones; several wetland areas; the Chindwin confluence: Adapted from Charles-Robin Gruel, Jean-Paul Bravard, and Yanni Gunnell, *Geomorphology of the Ayeyarwady River, Myanmar: A Survey Based on Rapid Assessment Methods* (Washington, DC: World Wildlife Fund, 2016), figures 45, 9, 88, 37–38. Drawn by Bill Nelson.

Map of the Ayeyarwady Delta: Drawn by Bill Nelson.

The Ayeyarwady floodplain: Reproduced with permission from Charles-Robin Gruel, Jean-Paul Bravard, and Yanni Gunnell, *Geomorphology of the Ayeyarwady River, Myanmar: A Survey Based on Rapid Assessment Methods* (Washington, DC: World Wildlife Fund, 2016), figure 61.

Rainfall pattern in Burma: Adapted from T. Furuichi, Z. Win, and R. J. Wasson, "Discharge and Suspended Sediment Transport in the Ayeyarwady River, Myanmar: Centennial and Decadal Changes," *Hydrological Processes* 23 (2009), 1631–41, figure 2, https://doi.org/10.1002/hyp.7295. Copyright © 2009 John Wiley & Sons, Ltd.

Alluvial islands: Drawn by Bill Nelson.

Method to determine ownership of newly created alluvial islands: Adapted from Benoit Ivars, "More Than Rice: A Contribution to the Ethnographic History on Resource and Frontier-Making in the Ayeyarwady Delta" (PhD diss., University of Cologne, 2022), figure 23. Drawn by Bill Nelson.

CHAPTER FOUR Intervention

Total forest loss; embankments in the Ayeyarwady Delta region: Adapted from Charles-Robin Gruel, Jean-Paul Bravard, and Yanni Gunnell, *Geomorphology of the Ayeyarwady River, Myanmar: A Survey Based on Rapid Assessment Methods* (Washington, DC: World Wildlife Fund, 2016), figures 12, 23. Drawn by Bill Nelson.

Pumped irrigation areas; location of hydropower dams; mining activity: Adapted from Ayeyarwady Integrated River Basin Management (AIRBM) Project, *Ayeyarwady SOBA 2017: Synthesis Report, State of the Basin Assessment,* vol. 1 (Yangon: Hydro-Informatics Centre, 2017), figures 9.9, 6.12, 4.8. Copyright © Myanmar Information Management Unit 2024. Used with permission. Drawn by Bill Nelson.

Model of village, cropland, levees, and delta: Drawn by Bill Nelson.

Gold mining: Photo courtesy of Radio Free Asia.

Sand mining: Photo by Sumaira Abdulali / CC BY-SA.

Insecticide risk level: Reproduced from Ayeyarwady Integrated River Basin Management (AIRBM) Project, *Ayeyarwady SOBA 2017: Synthesis Report, State of the Basin Assessment,* vol. 1 (Yangon: Hydro-Informatics Centre, 2017), table 4.3. Copyright © Myanmar Information Management Unit 2024. Used with permission.

CHAPTER FIVE Nonhuman Species

Ayeyarwady river dolphin: Photo by Stefan Brending, CC BY-SA 3.0 DE.

Snow carp: Photo courtesy of Xiao-Yong Chen. Figure 2 from Tao Qin et al., "Five Newly Recorded Cyprinid Fish (Teleostei: Cypriniformes) in Myanmar," *Zoological Research* 38, no. 5 (2017): 300–309. https:/doi.org/10.24272/j.issn.2095-8137.2017.063.

Hilsa: Photo by oqba/stock.adobe.com.

Copepods: Photo by Andrei Savitsky, CC BY-SA 4.0.

White ginger: Photo courtesy of Jana Leong-Škorničková.

Mollusc: Photo courtesy of Dr. Ekaterina S. Konopleva.

Oriental darter: Photo by Kaippally, CC BY-SA 3.0.

Burmese roofed or smiling turtle: From John Anderson, *Anatomical and Zoological Researches: Comprising an Account of the Zoological Results of the Two Expeditions to Western Yunnan in 1868 and 1875; and a Monograph of the Two Cetacean Genera,* Platanista *and* Orcella, vol. 2: *Plates* (London: B. Quaritch, 1878), plate LXVIII; image posted by Philbert Charles Berjeau to the Flickr Biodiversity Heritage Library.

Asian small-clawed otter: Photo by ArtMechanic, CC BY-SA 3.0.

CHAPTER SIX Iatrogenic Effects

Mozambique tilapia: Photo by Greg Hume, CC BY-SA 3.0.

Giant gourami: Photo by Joshua Sherurcij.

INDEX

Page numbers in italics indicate illustrations.

accretion vs. avulsion, xiv–xv, 31
agriculture: on alluvial islands, 104; cereal agriculture (grains), 154, 157–59, 178–79; in China, 70; commodities vs. diversity, 163, 164–66; deforestation, 36, 66, 74–75, 160; domestication as practice, 153–54, 155–56, 157–58, 164–66; drainage and diking for, 74–75, 121–22; in Dry Zone, 55; early irrigation methods, 55; fertilizer runoff, 130; fixed-field agriculture, 65–66, 107, 154; "flood recession agriculture" method, 54, 157; flood-retreat cultivation, 100; impact in general, 153–54, 156–57, 160–62; impact on rivers, 53, 73–74, 75–76, 107–8, 140; impoverishment of soils, 75; insecticides and pesticides, *131*, 131–32, 147; repeated annual planting, 159–60; replacement of wild nature for, 155–56; seeds for, 158
Ain Ghazal (town), 68

alluvial islands, 103–6, *106*
Amazon River, 15, 47
animal farming and production, 163, 164–66, 177
Anthropocene: thick, 53, 65, 189n1; thin, 53–54, 65, 189n1
anthropocentric view: of landscape and nature, 161–63, 184; of rivers, xii–xiii, 4
antibiotic resistance, 177, 178, 181
aquaculture, and domesticates, 166–67
Asian hairy-nosed otter (*Lutra sumatrana*), 149–51
Asian small-clawed otter (*Aonyx cinereus*), 149–51, *150*
Aung Thin, U, 84–85
avulsion vs. accretion, xiv–xv, 31
Ayeyarwady Basin (Burma): ecological zones, *93*; endangered forest types, 119, *120–21*, 121; habitat loss, 119, 148, 151; hydroelectric dams, *124*; pumped irrigation, *122*, 122–23

213

Ayeyarwady Delta (Burma): alluvial islands, 104; embankments, levees, and dikes, 123, *125*, 125–26, *126*; formation, 101; head of and distributaries, 93–94; map, *94*; as tidal zone, 96

Ayeyarwady River (Burma): alluvial islands, *103*, 103–6, *106*; beginning of, 91; channel migration, 102; confluence (Myitsone), 81–84, 91; course change, 16, 21–27, 90, 102; cultural connection through water, 59–60, 203n18; dead zones, 130–31; elevation levels, *89*, 94–96; fish decline, 110–17; floodplain of, 54–55, *95*, *96*, *97*, 119; flood-prevention measures, 126, *126*, 181; flood pulse of, *42–43*, 90, 91, 98–100, 101; flood-retreat fish capture, 100; geomorphology, xviii; gradient, 94, 97, 98; hydroelectricity, 123, 139, 173; industrial-scale transformations, 109, 111, 116–32; lateral shifts, *42–43*; map, *24*, *60*; as navigable river, 58–60, 195n25; navigation in high-water months, 26–27; navigation when shallow, 21–26, *26*, *27*; pilots for navigation, 24–25; pollution from industry, 130–32, 145, 147, 175; pollution from mining and fuel oil, 116–17, 127, 137; research collaboration on, xvii–xviii, 78, 197n1; seasonality, 97–99; sediment pulses, 101–2, 103; shape, 102; tributaries, *89*, 91–93, 123; upstream flow, 97–98; water volume, 97. *See also* mining (near Ayeyarwady River); *nats* (river spirits); watershed of Ayeyarwady River

Ayeyarwady river dolphin (*Orcaella brevirostris*), *133*; and tradition, 112; as voice for Ayeyarwady River, 134–36, 140–42, 145, 147, 149, 151–52, 200n1

bacteria: growth, 204n24; as problem, 177–78, 181, 205n25

basin of Ayeyarwady River. *See* Ayeyarwady Basin (Burma)

beavers, and change, 36

biodiversity, 41, 45, 163–64, 184–85

biomimicry, examples of, 50

birds, 48, 51, 146–47

Braudel, Fernand, 58

Brenz River watershed (Germany), *28*, 28

British colonial era, freshwater fleet on Ayeyarwady River, 59

Bügük Menderes River (Turkey), 29

Burma: cultural and social connection through water, 59–60, 203n18; Dry Zone, 55, *56*, 118, 119, 121; population growth, 118, 119; rainfall pattern, 97, *98*; situation in 2023, xv, xvi; use of name, xvi–xvii; visits and research by author, x–xi, xv–xvi, xvii–xviii; water extraction, 185

Burmans, 55, 59–60. *See also Burmans by name*

Burmese roofed or smiling turtle (*Batagur trivittata*), 147–49, *148*

canals for navigation, 170–71, *172*

catfish, 86, 113–14

cereal agriculture (grain growing), 154, 157–59, 178–79

channel migration of Ayeyarwady River, 102

charcoal, 66, 67

chicken, 164, 165

Chindwin River (Burma), 92, 97, *99*, *128*, 173

Chindwin watershed, 90

chlorpyrifos, 147

Cioc, Marc, 168, 171

climate change, 16

coastal plains, and the movement of rivers, 17–18, *18*, 19

commodities, and natural resouces, 161–63, 164–66
connectivity: afforded by water, 28–29; as concept and continuum, 45. *See also* culture, connection through water
copepods, *141*
culture, connection through water, 59–60, 203n18

Daly, Herman, 204n24
Danube River, 36, 39
dead zones, 130–31
deforestation: for agriculture, 36, 66, 74–75, 160; along Ayeyarwady River, 117–18, *118*, 119, 121; impact, 70–71; along rivers, 67–70, *69*, 170; for wood, 36, 66–68
delta of Ayeyarwady River. *See* Ayeyarwady Delta (Burma)
dikes. *See* embankments on Ayeyarwady River
disturbance, importance and role in ecology, 41–43. *See also* flood pulse
dolphin. *See* Ayeyarwady river dolphin
domesticates: and aquaculture, 166–67; and biodiversity, 163–64; change in species, 164; metaphysical view, 167–68; as preferred species, 155–56, 157, 158–59, 167–68; vs. wild flesh in the world, 163–64
domestication: as practice in agriculture, 153–54, 155–56, 157–58, 164–66; as process, 163, 165–66, 167, 168; of rivers, 168–72, 176
Dry Zone, 55, *56*, 95, 118, 119, 121

economy vs. nature, 161–63
embankments on Ayeyarwady River, 123, *125*, 125–26, *126*
engineering of rivers: by humans, 108–9, 169, 178, 184; iatrogenic effects, 177–78, 179–81; as repurposing, 109, 171, 172, 173–74, 185–86; soft-path engineering as solution, 186–88
The Epic of Gilgamesh, 68
Erie Canal, 172
European beech (*Fagus sylvatica*), 13
Everest, George, 35
exponential growth, 204n24

fall line, as border, 57, 61
fauna, and river in its entirety, 5. *See also* species
fire: as disturbance, 41; as tool, 64–65, 182
firewood, 66, 67, 68, 162
fish: decline in Ayeyarwady River, 110–17; and flood pulse, 39–40, 46–47, 48; food of, 141–42; migrations, 47–49, 51, 111–12, 139; view of river today, 136–40
fish farming, domesticates in, 166–67
fishing in Burma: illegal fishing, 113; species targeted, 113–14; techniques, 100, 111, 112–15, 137, 138
Fisk, Harold, 20, 21
floating cargo, 61–62, *62*, 195n25
floodplain of Ayeyarwady River: early agriculturists, 54–55; map, *95*; wetland areas, *96*, *97*, 119
floodplains: biodiversity in, 41, 45; changes by humans, 36; and domesticated river, 172; drainage, 74; early humans and city-states, 52–54; importance of floods, 38–39, 42–43; movement of water, 43–44; restoration, 187–88; below riverbed, 20
flood pulse: for agriculture, 55; as cycle for the movement of life, 45–49; description, 38, *44*; and fish, 39–40, 46–47, 48; importance and role, 38–39, 41, 45–46, 48, 74; interspecies-assisted movement,

flood pulse (continued)
49–51; and life-forms, 40–41, 45–46, 48–49, 99–100; and sediment, 101; as term, 40, 101

flood pulse of Ayeyarwady River: description, 90, 91; lateral shifts, 42–43; and life-forms, 99–100; and monsoon, 90; and sediment, 101; trajectory, 98–99

"flood recession agriculture" method, 54, 157

flood-retreat cultivation and fish capture, 100

floods: change to lines on maps, 15; damage from, 182; humans as cause, 176–77; hundred-year floods, 34; importance and role, 38–39, 42–43, 44, 182; and life-forms, 39, 46; microbial productivity, 45; probabilities and data, 34–35, 37; protection of humans (flood prevention), 126, *126*, 180–82; receding waters, 43–44; risk monitoring, 191n13

flow of river: and change, 27; and meandering river, 30–33; in upstream Ayeyarwady River, 97–98

forestry, goal and impact, 161–62, 188

forests: as commodities, 161–62, 164; humans' view of, 160; single-species forests, 162. *See also* deforestation

forest types in Ayeyarwady Basin, endangered status of, 119, *120–21*, 121

freshwater mussels, movement and reproduction, 49–51

friction of distance, 57–58

"galactic time," 8–9

Gallus gallus (chicken), 164. *See also* chicken

Ganges River, changes in and maps of, 35–36

geological time, 2–3, 9, 16, 89–90

geomorphology of Ayeyarwady River, research on, xviii

Germany, forestry in, 161–62, 188

giant gourami (*Osphronemus goramy*), 166–67, *168*

glacial and interglacial periods, 9, 11, *11*, 12, *13*, 16

glochidia (larval mussels), 50

gold mining, 116, *127*, 127–28

Gorky, Maxim, 156

grain growing (cereal agriculture), 154, 157–59, 178–79

gravel mining, 129, 143–44

"Great Drying," 74–75, 122, 179, 181

habitat loss, 119, 148, 151, 184–85

hilsa (*Tenualosa ilisha*), *138*, 138–40

Hinthada, 93, 95

Homo sapiens: as domesticate, 164; as invasive species, 184–85; time of, 11–12. *See also* humans

Htiyaing, decline in fish, 110–12

humans: changes to rivers, 7, 36–37, 64, 65–67, 72, 107–9, 169, 173, 176–77, 179–80; engineering of landscape and rivers, 108–10, 169, 171, 172–74, 178, 184, 185–86; human waste, 174; as hunter-gatherers, 51–53, 64–65, 66, 153, 168–69; lifetime as temporal lens, 13–14; population growth, *72*, 72–73, 107, 118, 119; rivers as integrative corridors, 59–60, 203n18; sedentism, 53–57, 65–67, 100, 154; slave trade, 63; subduing of nature, 71–72, 108–9. *See also Homo sapiens*

hunter-gatherers: early humans and rivers, 52–53, 64–65, 168–69; impact on rivers, 64–65, 153; patterns of life, 51–52; wood use, 66

INDEX

hydroelectric power and dams, 173, 174, 187; and Ayeyarwady River and Basin, 123, *124*, 139, 173
hydrological system, and connectivity of water, 28–29
hydrology: in repurposing of rivers, 109; research on Ayeyarwady River, xvii–xviii

"iatrogenic," as term, 175–76
iatrogenic effects: examples, 177–79, 181–82, 205n25; of river engineering, 177–78, 179–81
iatrogenic illness, 177, 179, 182–83
industrialization, 108
industrial pollution, 130–32, 145, 147, 175
industrial-scale transformations and the Ayeyarwady River: description, 109, 117–18; embankments and irrigation, 121–26; and fish decline, 111, 116–17; forests and deforestation, 117–21; hydroelectric dams, 123; mining, 128–29; pollution, 127–28, 130–32
insecticides and pesticides, *131*, 131–32, 147
Irrawaddy Flotilla Company, 59
irrigation: early methods, 55; impact, 74–75, 121–23, 140, 174; mistaken views, 54; pumped irrigation in Ayeyarwady Basin, *122*, 122–23
Ivars, Benoit, 104, 105

jars as raft, 61, *62*, 63

Kaifeng (China), and course of Yellow River, 19, 20, 180
Ko Maung Maung Oo. *See* Maung Maung Oo

Landis, Abbie Gascho, 50
larval mussels (glochidia), 50

law for boundaries (accretion vs. avulsion), xiv–xv
levees. *See* embankments on Ayeyarwady River
life-forms (nonhuman species): and dead zones, 131; exclusion by humans, 134–35, 140–41, 151–52; and flood pulse, 40–41, 45–46, 48–49, 99–100; and floods, 39, 46; impact of humans on, 72–73, 75–76, 184–85, 206n32; impact on rivers and watersheds, 108; and insecticides, *131*, 132, 147; and mining, 136–37, 143–45; poaching and trafficking, 150–51; and pollutants, 175; and rivers, 4, 5, 77; as voices for Ayeyarwady River, 134–36, 140–42, 145, 147, 149, 151–52, 200n1

Maikha river (or N'mai Kha), 91, 173
Mali Kha river (or Malikha), 91, 173
mangrove forests, felling of, 119
maps: and change, 35; rivers as line, 3, 15, 16, 21; and "river time," 21
Maung Maung Oo, 78, 113–17, 197n1
"meander," as term and pattern, 29
meandering river: and connectivity, 45; diagram, *32*; examples, 21, 22–23, *33*; factors and measure, 29–30, *30*; and flow of water, 30–33; irregularity in, 31–32; vs. straight-line river, 29, 30, *30*, 170–71
"meanderness," 29, *30*
meanders, as movement, 29–33
Mencius, 71
metallurgy, 67
migration: of fish, 47–49, 51, 111–12, 139; pattern and role, 52–53; of species, 51–52
mining (near Ayeyarwady River): activity in Ayeyarwady and Chindwin watersheds, *128*; gold mining, 116, *127*, 127–28; impact,

mining (near Ayeyarwady River) (continued)
   116, 127–28, 129; impact on life-forms, 136–37, 143–45; revenue from, 128–29; of sand and gravel, 129, *130*, 143–44; sediment from, 116, 129, 137, 144–45
Mississippi River: changes in, 20–21, 22–23; floating cargo, 195n25; floodplain, 39; floods, 48, 182; "meander belt," 21, 22–23; movement, 14–15
mollusc (*Lamellidens mainwaringi*), 144, 144–45
monsoon, 90, 97, 126
movement: temporal lens, 8–14; universality, 8–9
movement of rivers: anecdote about Penn's Creek, xiii–xiv; blocking of own route and "jumping" to a new bed, 17–20, *18*, *19*; and climate change, 16; data for (time-series data), 34, 35; domestication of, 169; as essence of rivers, 15, 33–34; external factors, 15–16; lateral movement, 45–46; meanders as, 29–33; movement by own action, 17–20; present status and the future, 35; rivers as movement, 14–20; and sediment, 17–18, 19–20, 69–70; vs. stasis, as presumption, 34, 37
Mozambique tilapia (*Oreochromis mossambicus*), 166, *167*
Mu River, 92
mussels, movement and reproduction, 49–51
Myanmar, use of term, xvi–xvii. *See also* Burma
Myitnge (little river), 92
Myitsone (confluence), 81–84, 91

Naing Tun Lin, 78, 110–13, 115, 197n1
nation-states, in repurposing of rivers, 109. *See also* state

nats (river spirits): categories of, 80–81; at confluence of Ayeyarwady River, 81–84; description, role, and importance, 77–78, 81, 197n1; festivals, 87–88; fish as, 86; migrant *nats*, 85–88; offerings to, 83–86, 112, 113; origin and list of, 78–81; patterns of worship, 84–85; research as collaboration, xvii–xviii, 78, 197n1
nature: replacement for agriculture, 155–56; subduing by humans, 71–72, 108–9; utilitarian view, 161–63, 184; as wasteland, 160
navigable water: Ayeyarwady River as example, 58–60; as contact zones, 58; sculpting of river for, 170–71, 172; and trade, 56–58, 61–63, 170; as transportation means, 57–58, 61–62
nga-ywe (catfish), 86
N'mai Kha river (or Maikha), 91, 173
nonhuman species. *See* life-forms
Nowitna River (Alaska), 33

oriental darter (*Anhinga melanogaster*), 145–47, *146*
Orissa (India), floods in, 176–77
overland transport, vs. by water, 57–58
oxbow and oxbow lake, *32*, 32–33

Pacific Northwest of North America, early inhabitants, 53
Pangaea, movement of tectonic plates, 9, *10*
Pan Jixun, 169
Penn's Creek (Pennsylvania), movement in, xiii–xiv
Peru, potato farming, 163
pesticides and insecticides, *131*, 131–32, 147
pigs, as meat, 165–66

INDEX 219

plants: aquatic plants, 143–44, 184; biodiversity loss, 184–85; cereal agriculture, 154, 157–59, 178–79; domesticates, 157–58, 159, 164; floods and flood pulse, 46; fungal diseases, 178; impact of humans, 143–44; infectious diseases, 159, 160; propagation, 158; and river in its entirety, 5; species identified, 157; in understanding of a river, 5, 6–7. *See also* species
Ploeg, Jan Douwe van der, 162–63
point bar, description of, 31
pollution on Ayeyarwady River, 116–17, 127–28, 130–32, 137, 145, 147, 175
potato, and landscape, 163
pottery and pots, 67
Pyu people, 55, 93

rafts, 61, *62*, *63*, 67
rainfall pattern in Burma, 97, *98*
Reice, Seth, 44
restoration, of floodplains and wetlands, 187–88
Rhine River, 34, 171
riverbed, level above floodplain, 20
rivers: beginning and formation, 2–3; connectivity of water, 28–29; human-caused change, 7, 36–37, 64, 65–67, 72, 107–9, 169, 173, 176–77, 179–80; and life-forms, 4, 5, 77; life history of, 1–2; as line on maps, 3, 15, 16, 21; main stem or channel as understanding, 5, 27–28; perspectives on, 3–4, 5–6; "river time," 2, 20–27; source of, 4–5; as system or whole in view, 5–7; terms for control by humans, 169; tributaries and distributaries omitted in understanding, 5, 6, 28; watershed as understanding, 6–7, 27–29. *See also specific topics*

sand and gravel mining, 129, *130*, 143–44
Schiller, Friedrich, 171
Schwenk, Theodor, 27
sedentism of humans, 53–57, 65–67, 100, 154
sediment: and alluvial islands, 103; and meanders, 31, 33; from mining, 116, 129, 137, 144–45; and movement of rivers, 17–18, 19–20, 69–70; in Yellow River, 18, 19–20, 68, 69–70, 169
sediment pulses, of Ayeyarwady River, 101–2, 103
Shwekyetyet jetty and port, 116–17
Shweli River, 91
Sittaung River and watershed, 16, 90, 102
slavery, 63
slow horticulture, 153
Smil, Vaclav, 163
snow carp (*Tor yingjiangensis* or Tor barb), 136, 136–37
soft-path engineering as solution, 186–88
Southeast Asia, 89, 173
species: adaptation, 47; diversity, 91, 161, 162; and flood pulse, 49–51; interspecies-assisted movement, 49–51; migration, 51–52; perception of the world, 200n1; temporal lens and movement, 14. *See also* life-forms (nonhuman species)
state: and cereal agriculture, 154, 158–59, 178–79; modification of environment, 154–56; modification of rivers, 109, 178–79; protection against floods, 180–81
state-serving habitat, 155
stem (main stem) of river, 4–5
St. Lawrence River (Canada), formation, 3

straight-line river, vs. meandering river, 29, 30, *30*, 170–71
surface connectivity, 45
swamps and swamplands, 160–61

Tagaung, kingdom of, 79
tectonic plates, movement of, 9, *10*
tectonic time, 9, *10*, 11, 88–89
temporal lens: and change in rivers, 15–16, 17–18, 27, 90; geological time, 2–3, 9, 16, 89–90; human life as, 13–14; and movement, 8–14; "river time," 2, 20–27; from widest to narrowest, 8–14
Thanlwin river (or Salween), 197n4
thick Anthropocene, 53, 65, 189n1
thin Anthropocene, 53–54, 65, 189n1
time. *See* temporal lens
trade: transportation and value of goods, 57–58, 62–63, 170; by water vs. overland, 56–58, 61–62, 170
trees, movement of, 8, 12, *13*
tropical watercourses, 90–91
Tulla, Johann Gottfried, 171
Twain, Mark, 14–15

U Aung Thin, 84–85
Uexküll, Jakob von, 200n1
United States, 49–50, 160–61, 177, 191n13
utilitarian view: of landscape or nature, 161–63, 184; of rivers, xii–xiii, 4

vernacular riverscape, 108
vessels as cargo. *See* floating cargo

waste, downstream movement, 174–75
water extraction in Burma, 185
waterscape of river, for understanding rivers, 6, 28–29
watershed: example of Brenz River, 28, *28*; fish in, 47; impact of humans, 107–8; understanding as system for rivers, 6–7, 27–29
watershed of Ayeyarwady River: distributaries, 94; geology and geological time, 88–90; mining activity, *128*; and monsoon, 90, 97, 126; tributaries, *89*, 91–93, 123; westward shift, 90
wetlands: of Ayeyarwady floodplain, 96, 97, 119; draining of, 160–61; restoration, 187–88
white ginger (*Curcuma candida*), *142*, 143–44
windmills of Dutch, 155
wood (timber): deforestation for, 36, 66–68; transportation on river, 61; use of, 36, 66–67, 162
World Commission on Dams report (2000), 187

Yellow River (China): blocking of own route and "jumping" to a new bed, 17–20, *18*, *19*, 70; and deforestation, 68–70; domestication and reshaping, 169–70, 180; sediment, 18, 19–20, 68, 69–70, 169

YALE AGRARIAN STUDIES SERIES
James C. Scott, series editor

The Agrarian Studies Series at Yale University Press seeks to publish outstanding and original interdisciplinary work on agriculture and rural society—for any period, in any location. Works of daring that question existing paradigms and fill abstract categories with the lived experience of rural people are especially encouraged.
—James C. Scott, *Series Editor*

James C. Scott, *Seeing Like a State: How Certain Schemes to Improve the Human Condition Have Failed*
Steve Striffler, *Chicken: The Dangerous Transformation of America's Favorite Food*
James C. Scott, *The Art of Not Being Governed: An Anarchist History of Upland Southeast Asia*
Timothy Pachirat, *Every Twelve Seconds: Industrialized Slaughter and the Politics of Sight*
James C. Scott, *Against the Grain: A Deep History of the Earliest States*
Jamie Kreiner, *Legions of Pigs in the Early Medieval West*
Ruth Mostern, *The Yellow River: A Natural and Unnatural History*
Brian Lander, *The King's Harvest: A Political Ecology of China from the First Farmers to the First Empire*
Jo Guldi, *The Long Land War: The Global Struggle for Occupancy Rights*
Andrew S. Mathews, *Trees Are Shape Shifters: How Cultivation, Climate Change, and Disaster Create Landscapes*
Francesca Bray, Barbara Hahn, John Bosco Lourdusamy, and Tiago Saraiva, *Moving Crops and the Scales of History*
Deborah Valenze, *The Invention of Scarcity: Malthus and the Margins of History*
Brooks Lamb, *Love for the Land: Lessons from Farmers Who Persist in Place*
Jamie Sayen, *Children of the Northern Forest: Wild New England's History from Glaciers to Global Warming*
Michael R. Dove, *Hearsay Is Not Excluded: A History of Natural History*
Gregory M. Thaler, *Saving a Rainforest and Losing the World: Conservation and Displacement in the Global Tropics*
Lee Sessions, *Nature, Culture, and Race in Colonial Cuba*
Brian Donahue, *Slow Wood: Greener Building from Local Forests*
Merrill Baker-Medard, *Feminist Conservation: Politics and Power in Madagascar's Marine Commons*
James C. Scott, *In Praise of Floods: The Untamed River and the Life It Brings*

For a complete list of titles in the Yale Agrarian Studies Series, visit yalebooks.com/agrarian.